Educational and Technological Approaches to Renewable Energy

Walter Leal Filho/Julia Gottwald
(eds.)

Educational and Technological Approaches to Renewable Energy

PETER LANG
Internationaler Verlag der Wissenschaften

Bibliographic Information published by the Deutsche Nationalbibliothek
The Deutsche Nationalbibliothek lists this publication in the Deutsche
Nationalbibliografie; detailed bibliographic data is available in the internet at
http://dnb.d-nb.de.

Layout:
www.kumpernatz-bromann.de

ISBN 978-3-631-62264-3
© Peter Lang GmbH
Internationaler Verlag der Wissenschaften
Frankfurt am Main 2012
All rights reserved.

All parts of this publication are protected by copyright. Any
utilisation outside the strict limits of the copyright law, without
the permission of the publisher, is forbidden and liable to
prosecution. This applies in particular to reproductions,
translations, microfilming, and storage and processing in
electronic retrieval systems.

www.peterlang.de

Table of Contents

Preface 7

Part A

"Renewable Energies in the Light of Development Experiences
in Fifty Years, 1960-2010"
Nelson Amaro 11

"E-Learning: Sustainability, Environment and Renewable Energy in Latin
America: a Multinational Training Pilot Module at Postgraduate
Level"
N. Amaro, F. Buch, and J.B. Salgueirinho Osório de Andrade Guerra 41

The Challenge of Attracting High-Quality Technology Transfers to Non-
BRIC Countries: Chile and its Emerging Wind Energy Industry
A. Pueyo, M. Mendiluce, D. Morales, R. García 69

Fostering Renewable Energies in Small Developing Island States
Through Knowledge and Technology Transfer: Findings from a
Labour Market Survey Undertaken in Mauritius under the DIREKT
Project
*V. Schulte, D. Surroop, R. Mohee, P. Khadoo, W. Leal Filho,
J. Gottwald* 89

Defining a Mitigation Strategy in a Developing Country Context:
The Case of Chile
R. O'Ryan, M. Díaz, J. Clerc 105

A Methodological Proposal for Community Participation in the
Development of Microgrid Projects
N. Garrido, M. Álvarez, G. Jiménez-Estévez 127

Part B

Wind Power Scenario for Brazil
  N.J. de Castro, G.A. Dantas, A.L.S. Leite 145

Energy Recovery from Biodegradable Waste in the Grain Processing Industry
  J.K. Staniškis, I. Kliopova, V. Petraškienė 159

A Study of Voltage Dips and Disturbances in Spanish Photovoltaic Power Plants
  J. Guerrero-Pérez, F. Espín, J. Martínez, A. Molina-García,
  E. Gómez Lázaro.. 171

Renewable Energy Policies that Impact Climate Change
  – The Case for Photovoltaic Solar Technology
  Nasir J. Sheikh, Tugrul U. Daim 183

Obstacles for Renewable Energy and Energy Efficiency in Chile –
  A Case Study from Hospitals
  L. Muñoz del Campo 213

About the Authors... 225

Thematic Index 235

Preface

The search for the means to promote renewable energy is a matter of great international concern, not only due to the high prices of conventional fossil fuels, but also because of the negative impacts of CO2 emissions on the world's climate. Even though the theme "renewable energy" has been treated as a matter of marginal relevance in the past, it is a key issue in the present and a matter whose relevance is likely to increase in the future. The reasons for this are twofold.

Firstly, as the world population reaches the 7 billion mark, energy demands are expected to rise. Based on the forecasts on energy production, which seem to indicate that conventional fossil fuels will become less and less available, and – due to their progressive limitation – prices are likely to increase, there is a pressing need to look for alternatives to meet current and future energy needs,

Secondly, if we are to find alternatives to fossil fuels, we need to find effective means to produce energy from biomass, from the sun and wind. In this context, research on the one hand, but also concrete applications on the other, are greatly needed.

According to the International Energy Agency (EIA), the world energy consumption is projected to grow by 50 percent between 2005 and 2030. Due to the fact that less fossil fuels will be available to meet such needs, there seems inevitable that renewable energy sources will be used, to meet at least part of the growing demands for energy.

Against this background, HAW Hamburg has created a Competence Centre on Renewable Energy and Energy Efficiency (CC4E) and a Technology Transfer Centre on Renewable Energy, whose goals are to undertake research and projects aimed at fostering the cause of renewable energy, and use technology transfer as a tool to helping developing countries to meet their needs. And since there is a paucity of publications which specifically address matters related to renewable energy in developing countries, we thought a book on educational and technological would be a timely contribution to the international debate on the topic.

This book therefore documents and disseminates a number of educational and technological approaches to renewable energy, with a special emphasis to European and Latin American experiences, but also with experiences from other parts of the world. It was prepared as part of the project JELARE (Joint European-Latin American Universities Renewable Energy Project), undertaken as part of the ALFA III Programme of the European Commission and involving countries in Latin America (e.g. Bolivia, Brazil, Chile, Guatemala) and in Europe (Germany and Latvia). Thanks to its approach and structure, this book will prove useful to all those active in the development of the renewable energy sector, especially those concerned with the problems posed by lack of expertise and lack of training in this important field.

A word of thanks goes to all authors who have contributed to this volume, as well as to all JELARE project partners, who made the project such a great success. It is hoped that this book will catalyse the development of further educational approaches in the field of renewable energy, and encourage their use in implementing new technologies. Enjoy your reading.

Walter Leal Filho & Julia Gottwald
Winter 2011/2012

Part A

"Renewable Energies in the Light of Development Experiences in Fifty Years, 1960-2010"

Nelson Amaro[1]

Abstract

Political, socio-economic and environmental trends are examined in the past fifty years. Three periods are distinguished in this time span. The first one is the "optimistic" phase (1960-70). Concerns about renewable energy were absent. The motto here is "development without any frontier". The second phase is the "pessimistic" stage (1970-85), where "the limits of growth" are emphasised. Interest in renewable energy is strongly brought to the fore at this stage. An environmental catastrophe is predicted if development patterns continue. Renewable energy becomes a viable alternative to expensive and contaminating fuel energy during this stage. The final phase, which we call "realistic", is being witnessed now (1985-present) where attempts are being made to reconcile development and environmental goals. These trends help to distinguish four paradigms that have oriented global development and renewable energy in the past sixty years: the "Modernisation" and "Neo-liberalism" school, which contributes to the optimistic vision of the sixties; Secondly "Dependence" theories followed by "World-Systems" schools, less concerned with renewable energies but looking at oil predominance as an instrument of big corporations and something serving the interests of rich countries. The "Club of Rome" paradigm, on the other hand, emphasises scepticism about all kinds of development efforts. In the "pessimistic stage" it predicted catastrophe if exploitation patterns continued without regard to environmental and clean energy concerns. The prevalent paradigm nowadays, however, is the "Sustainable Development" approach, which seems to be a synthesis of past experiences amenable to the "realistic" stage. This realisation will help to build bridges among extremist ideologies that continue defending the "development at all costs" that many proclaimed in the seventies. Universities

---

[1] The author is grateful to the European Union Alpha III Programme and the JELARE Project financed by it, led by the Hamburg University of Applied Sciences, for its continuous help and support. N. Amaro is with the Galileo University in Guatemala, 7ª. Avenida y calle Dr. Eduardo Suger Cofiño, Zona 10, Guatemala, CA. (E-mail: nelsonamaro@galileo.edu).

may play an objective role in favour of renewable energies at this point in time. This effort might become an important contribution to the 450 Scenario endorsed by the International Energy Agency, which envisages limiting the global temperature rise to 2°C above pre-industrial levels by the year 2030.

Latin America and the world have experienced big swings from the "First" to the "Fifth Development Decade" (1960-2010), following the denomination coined by the United Nations (UN). Public policies had initiatives with ups and downs similar to the major trends of the time. The oil crisis reached it maximum point in 2008, when its price reached US$147 in July. This event, unique in the history of fuel, immediately led to a series of measures to achieve more energy efficiency. The responsible bodies of many countries made energy production matrixes that, for the future, presented a gradual reduction of fossil fuels in favour of different renewable energy alternatives.

This effort still needs more time to be evaluated, but it is adequate to appraise it in the light of the development context where it has taken place. The crucial question for the future is: will past patterns continue into the present, or will the contexts in which this situation has emerged change sufficiently to produce new results? Since 1960 the rise in oil prices has determined most initiatives in renewable energies, which have gone forward as high costs have prevailed. Experience shows that when the price of oil has declined, efforts to design, boost, invest and produce these energies lose impetus.

In this document, we will analyse the different development contexts that have taken place in the last 50 years. We will make the paradigms that have influenced this result clear, determine its impact in the dilemma of fossil fuels-pollution versus renewable energies, and infer from this analysis the probable course of the trend, in order to derive lessons in sustainable development for the present and the future, – especially with regard to the role that universities may play in this dilemma.[1]

I. Background

The 1960s had the "blessing" of the world community, calling this period "The First Development Decade". This "baptism" does not mean that inequality among nations was not examined by classical theoreticians as far back as the 19th and early 20th centuries (e.g. Adam Smith, David Ricardo, Auguste Comte, Herbert Spencer, Karl Marx and Max Weber to name only a few). Nevertheless, less favoured or less developed nations were regarded more as simple societies that had not yet undergone sufficient evolution, or simply as objects of the colonial and imperialist policies of the most advanced capitalist countries. Also, these thinkers did not use the concept of "development" as such for their analysis. They referred to similar processes as examples of "evolution", probably influenced by Charles Darwin.[2]

By the period immediately after the Second World War nations were gradually beginning to find consensus on a vision of development. This vision, as we know it today, starts to be legitimate in the period after the foundation of the United Nations, the creation of the Marshall Plan for the recovery of Europe (1947-51), and the creation of President Truman's Point Four Program (1949). (The latter preceded incidentally the creation of the United States Agency for International Development [USAID] in the sixties, which would also take part in the stages to come.)

However, while this profile was being created by the decision-makers at the top, the popular image of the developing world before World War II might be inferred from the famous Tarzan movies of the time, the novels of Edgar Rice Burroughs or even, more recently, the Hollywood adventures of Harrison Ford as Indiana Jones. In order to emphasise the public perception of the changes to come in the sixties, it is important to look at a few nostalgic aspects from the fifties. It is looked at as a "time of innocence". Music, some movies and documentaries, social centres with gramophones and "big band" music, drive-ins, cafeterias with decorations from this epoch etc, are a clear sign that social practices had a sudden change, and that these former times were yearned for.

This evolution and the way public policies are conceived in the "*Development Decades*" can be summarised, after these precedents, in 3 phases: one might be called "optimistic", the second "pessimistic" and the third – still facing us today – might be labelled "realistic". Next, we will set out the characteristics of each phase and examine the role played by efforts to promote renewable energies in this context.

II. Development Phases

A. Optimistic Phase

Spanning from 1960 to the beginning of 1970, this is the "First Decade of Development", as the United Nations System called it. A series of singular events point towards changing times:

- The independence of the African countries from their colonial rulers.
- The promise made by the most advanced countries (confirmed in United Nations conclaves) to help developing countries with 0.7% of GNP. (This

goal was incidentally resurrected by world heads of state evaluating the Millennium Objectives in New York at the end of 2010.)
- Defence of civil rights together with generational and student protests represented by the "hippie" movement.
- The "May Movement" in 1969 in France, under the motto "Imagination takes Power", which challenged the Establishment and kept the country in a constant state of agitation with street marches and confrontations with the security authorities.
- De-Stalinisation in the Soviet Union, promoted by the 20th Congress of the Communist Party in the fifties and sixties, when the main guidelines of the Stalinist period were criticised and rejected.
- The guerrilla movements in Latin America, spurred by Che Guevara's call to instigate a series of "Vietnams" on the continent.
- The call for reforms from the Second Vatican Council.
- The unique influence of the coincidence of reformist personalities in key decision-making positions around the world, such as John Kennedy, Nikita Khrushchev and John Paul II.

All these events anticipated a better future regardless of the different ideological approaches, and even though these events occurred in the context of a "Cold War" mentality.

1. The Role of Renewable Energy

Following the premises of "unlimited growth", oil was barely acknowledged as a strategic and non-renewable resource for industrial societies in general, or for the development of "emerging" countries, as they have been recently named. In addition, by all accounts, oil followed market laws, and – compared to recent trends – was extremely cheap. Big multinational entities controlled oil production, especially in Arab countries, and its major sources besides the Middle Eastern countries were the great powers: the USA and the Soviet Union. This started a confrontation after World War II that lasted for almost the remainder of the 20th Century.

The ideas through paradigms that influence this outcome will be examined later on. At this point it suffices to highlight that renewable energy or "energy alternatives" to oil hardly received any attention. The need was not felt. The puzzle that all parties wanted to solve at this stage was how to bridge the differences between developed and developing nations. In any event, during the sixties all the main players contributed to the prevailing mood: an optimism that promised a world without inequalities and a better future for everyone. This outlook was explicitly voiced in the United Nations' "Declaration on Social Progress and Development" (1969).[3] This document does not contain a single reference to the energy problem. As a result, this predominant vision did not include environmental concerns. Natural resources, including renewable energy sources, were regarded either as instruments of the colonial powers or merely as signs that the system had failed to exploit this wealth in favour of the poorest.

B. Pessimistic Phase

This approach and the spirit of the time started to change at the end of the sixties. The transition is described by Dumar Suárez as follows, depicting to a great extent the rise of the Organisation of Petroleum Exporting Countries, OPEC:

> "Until the early seventies, oil supply did not seem to constitute a problem, given that the demand grew almost in parallel with the discovery of new oil wells, and prices kept low... However, during that time, a slow but firm rise in prices started, and it became abrupt in 1973 and 1974. After that, it was soft again, and in 1979 it was again abrupt. (It is important to take into account that before, in 1972, the Suez Canal was blocked by the Yom Kippur War, forcing oil companies to go around Africa by the Cape of Good Hope, with the resulting increase in prices, which, along with the increase of 1973, created a panic environment in the stock markets of the world)."[4]

On the other hand, Kenneth Boulding's famous metaphor, which turned out to be prophetic, was precisely suggested when this change began. The "frontier" mindset characterises the optimistic attitude and its vision of the unlimited exploitation of resources and population growth, where movements may occur indefinitely. This mentality is about the conquest of nature, mastered by the intervention of mankind. There are no limitations for that possibility and for the satisfaction of human needs. So, growth is infinite and expansion has no borders.

This vision was coming to an end, just as Boulding predicted in 1969 near the end of the optimistic phase. The vision that came to replace the "frontier" mentality envisaged the whole planet as a spaceship. Earth with its inhabitants is on a long trip into a finite and fragile world. This spaceship is crewed by a population that must take into account the limitations of travel just like passengers on any ship. This image also implies that the available food, water, etc. in the ship is limited (i.e. non-renewable) and that its consumption, with the resulting waste that needs to be managed and pipelined, must therefore be planned and controlled to ensure the final destination is reached.[5]

The end of the optimism that characterised the previous phase can be attributed to many causes, some being more important than others. The following events form a non-exhaustive list:

- The aforementioned oil crisis.
- The recycling of the "petrodollars" suddenly captured by oil-exporting countries and channelled into the western financial and banking system, which were supplied, with facilities, to anyone that could show a certain credit capacity. This included sovereign countries with developing economies and weak fiscal restraints.
- The institutionalisation of "foreign debts", galloping as a consequence of the state of affairs outlined above, in national budgets.
- "Structural Adjustments" corresponding to advice derived from the "Washington Consensus" aiming to "put the house in order".[6]
- The spread of military and authoritarian regimes throughout the world.
- The unprecedented stagnation and inflation arising in the US and influencing the whole world.
- The rise of revolutionary armed movements either as a result of frustration after independence failed to materialise in the sixties, as in Mozambique or Angola, or insurgent movements as in Central America. In Nicaragua, following one such insurrection, a regime similar to the one in Cuba took power; the same happened in Chile through elections, although the regime was overthrown by a coup d'état in 1973. The regime in Nicaragua ended by a majority vote of citizens in 1990.
- Resistance to change by elites in many developing countries fearing being ousted from power.
- The investment contraction that followed the period of post-war prosperity.
- Macroeconomic imbalances and weaknesses in the promotion of the import substitution model of development, especially in Latin America.
- The combined effect of all these factors causes the labelling of this phase as "the Lost Decade" – a claim far from the optimism of the sixties.
- The rise of voices warning of the need to consider "the limits of growth" and the potential disaster caused by environmental erosion.

This situation forced policy and attitudinal changes in the approaches voiced by national and international institutions in charge of development.

The incorporation of *the foreign debt payment* in the annual budgets of most developing countries as a significant percentage, plus the growing influence of regional banks and the World Bank, with the resulting decline in the influence of UN specialised technical assistance organisms (UNESCO, WHO, ILO, UNFPA, etc.), characterises this period. The initial vision of entire organisations dedicated to promoting development becomes impossible when financial organisms with specialised technical branches give their support, donations and loans in the same technical areas that were previously reserved for these specialised international organisations. Gradually, multinational banks added technical skills that were previously the reserve of these specialised organisations. In turn, financial organisations expanded their scope beyond public finances, which had represented their main duties in the original UN design. This change weakened the action of these specialised international organisms and the UN as a whole.

On the other hand, it was at this stage that bilateral external aid started to reduce the amount allocated to development aid. Parliaments and congresses of donor countries started to cast doubt upon these expenditures. In particular, the illicit enrichment of many leaders in recipient countries contributed to this discomfort. The established goal of developed countries in the sixties of contributing 0.7% of GDP found significant opposition in the years afterwards, especially from countries with higher income. However, 50 years later some smaller countries in Europe, such as Holland and a few Scandinavian countries, reached this number and have even exceeded it.

These alarms took most Latin American and Caribbean countries by surprise. The role of the UN as a bridge between the countries of the northern hemisphere and Africa and Asia made their situation even more vulnerable. There were some "middle class" countries with more credit capacity, avid for resources that were translated into debt, with governments that quickly realised the inter-temporary inflexibility of budgets – especially in those items related to defence, salaries and purchases. Those who ignored this reality went into an inflationary spiral and experienced economic misfortunes that soon made them reconsider their expenditures. Public expenditure became the most important area for reform to counter treasury shortfall. The general public policy in this period was to control resources allocated in the social area, especially in education and health, in order to even the negative cash balance of inflation, current account deficits and increasing foreign payment debts.

Mexico's inability to honour its debt in 1982 was an alert for the whole region, and countries became aware of the necessary reforms. The international community shuddered in the mid-eighties when Carlos Andres Pérez, the President of Venezuela, embarked on a series of reforms in his second term that sparked a popular protest in which supermarkets were sacked and many people were killed. Those deaths were attributed to the repercussions of reforms on the poorest groups, especially the measures designed to put Venezuelan finances "in order".

Economic growth rates, encouraged at first by this sudden incorporation of resources, increased in the seventies, but started to fall in the eighties and were in fact dramatically reverted by the end of the decade. The spirit of the time was "pessimism" as opposed to the attitude prevailing in the sixties. To crown this trend, a claim was made that it had been a "lost decade" in Latin America and the Caribbean, instead of putting emphasis on the development goals actually reached. The crisis determined severe financial restrictions on public expenditure. Two options were presented: either tax collection was increased and/or public expenditure was reduced. The first option became virtually impossible due to the traditional tax evasion, investment discouragement and capital flight, especially in a recessive period.[7] The remedy would have been worse than the disease. Therefore, a reduction in public expenditure became imperative.

1. The Role of Renewable Energy

An enthusiasm for producing renewable energy starts to appear at this stage. Brazil is the best example. The coup d'état of 1964 started the first in a series of military regimes that lasted until 1985. Before the events of the seventies, one of the regime's top priorities was to accelerate the process of making Brazil one of the most developed countries in the world. Gradually these efforts were frustrated, to a great extent because of oil dependence. Nowadays, however, along with the USA, the country generates more than 70% of the ethanol produced worldwide, and its distribution throughout the world is part of Brazilian foreign policy. Today, despite the problems in the nineties, most vehicles in Brazil run on ethanol. Nevertheless, it is essential to acknowledge the origins of this effort. David Sandalow says:

> "The early 1970s were a boom time in Brazil, with many observers heralding the 'Brazilian economic miracle.' Yet President Ernesto Geisel faced twin problems. First, the cost of Brazil's oil imports tripled in late 1973, due to the Arab oil embargo. Second, world sugar prices, which had been climbing upward since the mid-1960s, declined sharply in 1974. Faced with these problems, Geisel launched the Brazilian National Alcohol Program in late 1975. The program was intended to reduce the need for oil imports and provide an additional market for Brazilian sugar. As a first step, the federal government immediately began promoting the production of ethanol for blending into gasoline, to the maximum extent feasible in existing vehicles (approximately 20% by volume) ... The results were dramatic. Between 1975 and 1979, ethanol production increased more than 500%."[8]

Nevertheless, alarmed voices were making themselves heard at the beginning of this period. Curiously enough, those voices were backed by university research and based on findings gathered by major scholars regarding the overexploitation of resources by humans. Simultaneously, international organisations began to use this research to encourage agreement among nations. Systematic approaches along these lines were articulated to develop true "paradigms", which eventually feed into the development phases described. Thomas Kuhn coined this term to describe the "puzzle", where the practice of science, far from being a uniform, gradual and accumulative process, takes different directions in the light of new premises.[9] In the next section the prevailing ideas of these periods will be examined and the different development "paradigms" influencing these events will be identified.

C. Present Situation: Realistic Phase

Voices demanding "reform with a human face" started to be heard in the mid-eighties. The Bolivian stabilisation programme, successfully carried out in 1985, offered a more realistic approach, especially because of its sensibility, through the "Emergency Social Fund", to the "poorest of the poor". Many countries worldwide approved the Social Investment Funds, and some other measures were enacted to alleviate the consequences of the reforms, building "safety nets" for the poor.[10]

Most countries in the area adopted the so-called "first generation" measures suggested by the Washington Consensus.[11] The "second generation" measures, which are still incomplete, are related to institutional strengthening (e.g. independent Central Banks, decentralisation, commissions around fiscal matters, justice reform aiming to reinforce the rule of law, educational reform, etc.). The launching of these policies at this stage allowed expansion into the next phase (1990 until now). Nevertheless, the implementation of these policies has been different in each country. The depth of reforms varies at this stage. These differences are showed in studies from the mid-nineties.[12]

The phase that we are experiencing now is a kind of synthesis integrating the previous two phases. The concept of "Sustainable Development" sums up the logic of these events. It is exactly a midpoint between the "optimism" of the sixties and the "pessimism" of the seventies. This phase refers to a growth with ecological limits, aiming at a temporary horizon that goes beyond one generation. There is a call for more pragmatism. Just as was pointed out in an international conclave at the beginning of the 21st century: "A more empirical pragmatic approach is needed".[13] Some factors that have influenced this transition are listed below:

- In Latin America, there is some economic recovery after the suffocation of the "Tequila Effect" that devastated Mexico in 1994, and the Asian crisis that affected, among others, Brazil in 1998. In both cases, the negative multiplicative effects that were expected did not materialise.
- Basic social indicators kept rising during the eighties, to a great extent due to inertia; the diffusion of technological development in health and access to medicines was similar, aside from governmental policies.
- As the 21st century continues, macroeconomic indicators in most countries are recovering. The "debt crisis" that many said was unaffordable has been reduced and become less of a real issue.
- Significant drops in the market, especially in technology and mainly in development centres like the USA, have been overcome since 2000, thus increasing incentive for investment in less developed countries.

- The measures taken by the Washington Consensus definitely helped achieve this result, although many have been keen to point out that these objectives have been achieved, to a great extent, at the expense of greater inequality in wealth distribution in most countries.
- However, the world financial crisis that we are facing today finds developing countries in a better position than in the nineties, in terms of macroeconomic indicators including debt.
- This all means that results have mainly been achieved by putting less emphasis on ideologies, or by adapting them to better correspond to reality. Two examples will suffice: the democratic coalition in Chile respected the macroeconomic decisions made during the time of Pinochet. Also, the flight of capital from financial centres that occurred immediately after the election of Lula in Brazil gradually ceased, and he proved to have a great respect for macroeconomic balances, which restored trust.
- Since the latter years of the last decade we have been facing a great paradox: concerns about debt and fiscal crisis have been transferred to the developed countries and no longer originate principally in developing or "emerging" countries.

1. The Role of Renewable Energy

At this stage, the availability of fossil fuels and their effects on the environment are a major concern in the world. Projections show that the availability of fossil fuels will be in danger in only a few decades and, consequently, prices will tend to increase.[14] In the past, this situation has promoted investment in renewable energy. Many countries have begun to produce energy matrixes for the long and medium-term, with quantitative goals where fossil fuel energy is gradually cut back, and renewable energies are gaining higher percentages over time.

A characteristic of the current period is that concerns over how to reconcile development and environment now extend over all society, and for the first time it is possible to detect a grassroots movement behind these propositions. Nevertheless, not all institutions have responded to the challenge at the same pace. Governments have been slow in tackling the questions of energy and mines exploitation. However, many governments have set up ministries and dependencies to take responsibility for the environment and the use of natural resources. In the case of Guatemala, for example, the Ministry of Environment and Natural Resources had a greater budget than the Ministry of Energy and Mines.

Active grassroots movements and international organisations have been faster in acting. Universities increasingly are adapting to the new situation because of demand not only from the market, but from government and pressure groups. But again, one sees greater effort put into environmental issues and natural resources, and less in energy and mines. As renewable energy is considered to come under the latter category, development tends to have a slower pace.

III. Paradigms of Each Phase of Development

The preceding sections have described the characteristics of development phases over time and highlighted how renewable energy has been perceived in each context. The next question to answer is the following: To what extent have the prevailing ideas in this context influenced these characteristics that have been outlined? Keynes says:

> "The ideas of economists and political philosophers, both when they are right and when they are wrong, are more powerful than is commonly understood. Indeed the world is ruled by little else. Practical men, who believe themselves to be quite exempt from any intellectual influence, are usually the slaves of some defunct economist."[15]

In this section, we will outline the main intellectual arguments and their roots, which have since influenced policy in the periods we have discussed. In doing so, we will follow Thomas Kuhn's concept of "paradigm", the "puzzle" in which every group in the scientific community helps produce new research that may eventually lead to "scientific revolutions". Thus the practice of science, far from being a uniform, gradual and accumulative process, takes different directions in the light of new premises, launching real thought-revolutions during specific periods.[16]

A. Modernisation Theoreticians

Many of the intellectuals who defined the pace of the sixties could be described as modernisation theoreticians. Walter Rostow was perhaps the most emblematic figure of the times, but the classic authors who helped launch and strengthen sociology as a science included Auguste Comte, Herbert Spencer, Emile Durkheim and others. What common elements can be highlighted between these authors?

- There is a common conception of progress as some sort of improvement, e.g. societies based on theological knowledge that grow to become metaphysic and scientific (Comte), or human groups that go from simple to complex composed societies (Spencer), or from mechanical to organic societies (Durkheim). Recently, a commonly mentioned trajectory has been that from traditional to modern societies (Parsons, Germani, Rostow, Lerner, McLelland, Hagen, Germani).[17]

- A more recent version is linked to the drastic fall of the socialist world, projecting images of a future without challenges, visualising the "end of history" – inferred from the work of Fukuyama, or the rescue of libertarian values proposed by Vargas Llosa, or Harrison and Rangel's outlines for adopting the right values of modernisation. All of them have a connection with the "neoliberal" movement that has defined the present time.[18]

- For example, Rostow, following the "social climate" of the sixties, tells us that development seemed to be around the corner. A "big push" was enough for the "traditional society" to reach, in consecutive stages, "The Age of High Mass Consumption", where material concerns gave way to different, less basic priorities, for example having kids (a reference to the "baby boom").[19] The development process, as some critics have pointed out, seemed to be a plane taking off on a one-way flight to the land of wealth!

- Cultural values or societies' intrinsic features produce changes in the economic sphere that are later transferred to family, education, politics, etc., for example "the achievement motivation" or innovations.

- Innovative businessmen and corporations with motivations such as religion or profit-lust, political arrangements, or any values in tune with the resultant changes introduce alterations that are later assimilated by entire national states.

- The main obstacles are related to values that are not suited for development, such as the tendency to enjoy leisure instead of having a work ethic, the right to "family" or "rentier" privileges, corporative ethics together with the presence of "wrong" religious values that reject technical and scientific approaches, especially those that are capable of transforming resources into market goods.

- The time needed to reach an era of abundance can be 40 to 150 years, judging by the example of more developed countries, mainly European countries and the USA. The agricultural stage in all these societies was very long, but in the industrial and service stages time periods become accelerated.

- Public policies emerge that faithfully follow the previous image. Economic aspects are the main concern and social aspects will be gradually solved. Investment rates are assigned to sectoral strategies, by-products of favourable opportunities in the markets. This approach seemed to be the primary mechanism for proceeding from one phase to another. Social development would generate the "trickle down" effect that would fight poverty, unemployment and marginalisation. Regional development policies are influenced by the concept of "growth poles". Support for innovative businessmen is focused on strategic sectors with explicit or implicit consequences for urban growth, middle-class emergence, orientation towards simple nuclear families and family planning, and the separation of state and religion.

- Their best choices were related to the diffusion of western institutional models into less developed countries. Also, "demographic transition" patterns behave according to the modernisation theoreticians and fertility has decreased following the declining pace of mortality. The popularisation of terms like "big push", "demonstration effect", "sectoral strategies", "dualism", "traditional and modern sectors" or also "post-modernism" may be traced back to this paradigm. The South East Asian countries, the so-called "tigers", follow this trend, perhaps giving more importance to savings rates, investments and education as innovation factors.

1. Position Before Renewable Energies

Their optimistic vision puts no limits on growth, and resources appear static or inert, just waiting for humans to generate wealth with them. That is why there is no vision of what "renewable energy" means for their suggestions, and why their vision fed "optimistic" decision-makers during the sixties. Any perception of the scarce availability of the natural resources needed for development, and the need to take care of these resources after being processed and turned to waste is non-existent and outside their mental framework. They resent the intervention of government in private development activities and generally discount assertions of the dangers attached to pollution and climate change. More extreme positions refuse to acknowledge any limitation to growth or any regulation or control of the exploitation of natural resources. Nonetheless, their position was more acceptable in the sixties, during the "optimistic" phase, than it is now.

A quick glance over the authors that represent this paradigm will show that universities in developed countries, particularly in the USA, are the most relevant proponents of these ideas, which to a great extent follow classic European authors. For example, Parsons, Rostow and Germani were directly or indirectly related to Harvard University; Lerner and Hagen were professors at the Massachussetts Institute of Technology (MIT). Fukuyama, probably the most recent of the modernisation theoreticians, is a graduate of Harvard and a professor at John Hopkins University.

B. The Paradigm of Dependency and the Club of Rome

Even if these two schools of thought are independent in their sources and evolution, they both emerged publicly during the stage that we are calling "pessimistic", at the end of the sixties and in the seventies. The "Dependency Theory" predicted that the stiffness of international stratification would prevent qualitative steps that could allow developing countries to raise their status to that of a developed country. The position of "dependent" countries was necessary so other countries could keep their dominance over them. This road did not lead to development.[20]

The Club of Rome announced that if the current trend of exploiting natural resources and waste management continued, Earth had only a survival capacity of 100 years. This statement meant that in the year 2070, life as we know it would disappear from the face of the Earth.[21] Even though there were differences between these two approaches, on certain points they held common ground: they were born contemporarily and they both announced that if current conditions continue, there would no longer be any possibility of development as we know it on this planet. This is the reason we decided to name them as precedents and influences on the "pessimistic" phase. They both have the following characteristics:

- The intellectual source and influence for both theories differs. The "Dependency Theory" has its foothold in the 19th and early 20th centuries through Marx, Engels and Lenin. Nevertheless, similarities can be found more recently in authors like Wallerstein, Arrighi and Frank.[22] From some perspectives, their ideas seem to have something in common with List and Sombart and the German School, because of their emphasis on national autonomy – although here the approach is more national than global. Antecedents of the Club of Rome can be tracked down to the 19th century, with Thomas Malthus and his work on population growth, including his famous statement about the opposition between the arithmetic growth of food and the geometric increase of population. This would inevitably lead to conflicts and wars as partial-control events, given that the projection would eventually become catastrophic – depending on the scale on which it occurs.
- The Dependency Theory, on the other hand, prioritises its focus on the structural and economic tension between countries in the centre and those on the "periphery", which produces internal and external alliances in order to perpetuate the situation of exploitation. The Club of Rome suggested a structural tension between development and environment, in which the first is pursued in such a way as to disqualify the second. As for the "dependency", the increasing income of transnational vehicles (direct investment, commerce, international cooperation and loans with the consequent external debt) produces

answers that contradict autonomy – they can, however, lead to a sort of precarious autonomy or to a step towards the semi-periphery, a kind of "in-between" state.
- Changes in the analysis of the environment, on the other hand, emphasise the "limits of growth" in such a way that pollution is reduced as the disappearance of non-renewable resources is acknowledged, "habitat" and population become balanced priorities, and industrialisation is moderated, favouring inputs with regulations and governmental control of exploitation activities. Both believe in stabilising action through government, which is rejected by modernisation theoreticians and neoliberals.
- Critics have voiced weaknesses in the conclusions of both theories. The "dependency theory" is, after all, one more dichotomy that is added to the traditional-modern or community society, or organic-mechanic solidarity classical scheme. It is now transformed into a new dichotomy: dependency-liberation (or autonomy). It is regrettable that there are only a few links to specific policies capable of transforming reality. On the other hand, two questions appear: is it inter-dependency or dependency? Why is it applied only to capitalists and not to socialists, when it is stated that there is no development without autonomy?[23] In its more recent World-Systems version, the generalisation is so wide that no link is allowed to specific policies in precise periods. With regard to this new approach, critics state that there is too much emphasis on globalisation without much empirical foundation. After all, the nation-state remains the most important thing, and remains the prevailing system.
- The paradigm of the Club of Rome has been criticised because it leaves some important variables out of its analysis. The evidence of economists, in particular, suggests that findings have not matched its predictions. Non-renewable resources, for example, can be replaced; prices can manipulate the use of resources if they start to be scarce; and alternative environment-friendly technologies can spring up, just as now with so-called "hybrid" cars. The arguments, their critics say, are more aimed at raising awareness, more ideological-political and less scientific, despite its objective appearance.[24] There is little empirical foundation to support the fears generated. For example, it is not the first time that the Earth's temperature has risen: the same thing happened in the Middle Ages, when there was no industrial society and therefore no consequent carbon dioxide (CO_2) emissions. Their proposed actions tend to immobilise economic development and call for more governmental intervention. This is one reason critics have suggested that, in view of socialism's failure elsewhere, discredited "reds" have now become "green" militants.
- In any case it is imperative to show the genuine insights of each theory. The Dependency Theory predicted the current dominance of multinational organi-

sations and international capitalist institutions. The Third World has never been so dependent on external capital flows and international monetary organisations. In its version of World Systems, the abandonment of the socialist system by many countries, led by the Soviet Union, was rightly forecasted. This version consistently predicted that the capitalist system had a worldly nature, and that those regimes were a version of the same just to gain influence. The theory is also pioneering in its forecasting of modern "globalisation", especially in its financial aspects. Its approach to the "underdevelopment of development" (Gunder Frank) is applicable in some areas, for example the north of Chile, where the development of saltpetre has been frustrated, or northern El Salvador, Eastern Guatemala and Western Honduras, where similar situations have arisen with indigo. These areas were engaged in development, which, due to external factors, turned into underdevelopment.

- Despite the critics, the influence of this work can be detected in governmental policies. Dependency Theory encourages a tendency towards economic autarchy with an economic, global, sectoral and vertical planning reinforcement. Also, it worsens tensions between its economic growth model and its social policies, frequently favouring permanent employment guarantees, high custom protectionist rates to defend national products, diversification of crops and exports, avoiding single buyers or importers in international trade. It favours the nationalisation of strategic industries with projections towards other productive areas and food self-sufficiency, stimulating economic flows control. Currently, it has shown a survival capacity in Latin America with the appearance of populist tendencies in countries like Nicaragua, Venezuela, Ecuador and Bolivia.

- With regard to the environmental version, ministries and policies aiming to preserve ecosystems and essential processes have been created in many countries. Action has been taken to preserve biological diversity and environmental conservation. It preconises environmental impact assessments as a prerequisite for authorisation to exploit natural resources. It protects transparency in the handling of information on this subject, should any action affect citizens. Four basic policies can be found: 1. Behaviour control by prices, specifically charging an amount for polluting fluids emission; 2. Pollution rights sale through permission for undesirable, yet controlled amounts; 3. Mandatory and control regulations; and 4. Promotion of renewable energies, although this is still subject to the ups and downs in oil prices.

1. The Role of Renewable Energy

We found these theories were most influential during the seventies and the first half of the eighties. In fact, during this period they raised awareness about the environment, and the issues that have more influence today were first mentioned during this period. Dependency Theory was more a critical series of clauses that challenged the defenders of modernisation theories within the context of the "Cold War". For example, Rostow saw his work as a "Non-Communist Manifesto". Also, Cardoso, probably its most outstanding author, abandoned the Dependency premises when he himself reached a position of political responsibility as President of Brazil. In some ways, energy resources were seen as part of the imperialist domination strategy by Dependency Theory, but a solid position on the use of these resources was lacking. Nevertheless, we should not forget that Brazil's emphasis on ethanol and renewable energy did not change during this period; indeed if anything it was reinforced.[25]

The Club of Rome, on the other hand, was completely focused on the subject and launched a dramatic call to the rest of the world and their intellectual decision-makers. In time, this position started to get more sophisticated and currently exhibits a greater appreciation of the social sciences. For example, there is more focus on economics, sociology and the political sciences in their arguments and proposals. There is consensus about the need to incorporate the environment variable as inherent to the development study processes. For instance, national accounts should incorporate the cost of environmental deterioration, and "environmental impact assessment" studies should be the rule and not the exception in all investment decisions, particularly when support is committed to sensitive areas such as the mining industry or the exploitation of natural reserves.

The Dependence Theory is, to a great extent, a response to the polarisation that occurred elsewhere in the seventies. Curiously enough, in contrast to past tendencies, this outlook has been developed by Latin American authors, attached to Latin American universities and international organisations (e.g. the Economic Commission for Latin America, ECLA, UN and FLACSO). Previously it was usually the case that centres of learning in developed countries were the producers of ideas, and developing countries consumed these ideas. This trend was reversed when this theory was accepted by many scholars from the USA and Europe. Nevertheless, this approach did not contemplate energy alternatives to the prevailing *status quo*. The academics' question was how to attain development, something that, in their opinion, was not possible under the influence of modernisation theories.

A more direct answer came from the Club of Rome. A group of scientists and politicians, including Nobel Prize winners, formed a coalition that gave the responsibility of researching the human impact on natural resources to the System Dynamics Laboratory of the Massachusetts Institute of Technology (MIT), under the direction of Dennis L. Meadows. Using quantitative methods and statistical models, the report already quoted – on the limits of growth – has proved to be one of the most influential works in more than three generations. The Club of Rome has now become a non-governmental organisation, and still influences public policies today.

C. The Paradigm of Sustainable Development

Gradually, the concept of "sustainability" has been introduced into development discussions since the second half of the 1980s. In 1972, during the Stockholm Conference, the United Nations Environment Programme (UNEP) was created. The change in the scale of environmental problems influences this approach, which turns this issue into a universal problem. Issues like rainfall volume and the influence of climate change have become a point of reference. This outcome might influence agricultural productivity, reduce or stop ocean flows, affect biodiversity or disseminate contagious diseases. The Greenhouse Effect (CO_2), the diffusion of toxic substances, soil acidification and acid rain and the reduction of the ozone layer (CFC) are part of these concerns. These threats contribute to the exaggeration of current perceptions regarding the security of energy supply, unequal access to energy across great swathes of population, and the problem of investment in an infrastructure capable of supplying energy.[26]

This recent evolution, however, anticipated the conclusion that, besides the theoretical positions that were being proclaimed, it was possible to reconcile development with environmental concerns. Development objectives need to be combined with the responsibility to leave a planet suitable to be inhabited and managed by future generations. As time passes, this paradigm becomes more coherent and, since it is proclaimed as a synthesis of past developments, its statements have become more fluid, demanding a kind of centrist position in a discourse where the paradigms of the last twenty years, reviewed above, still survive, and often appear extreme, sometimes polarised in this context.

At this point in the development story, in 1987, the meeting of the World Commission on Environment and Development was held, at which the Brundtland Commission provided an appropriate and timely definition of the concept of "sustainable development" after four years of work. This was defined as follows:

> "a process of change in which the exploitation of resources, the direction of investments, the orientation of technological development; and institutional change are all in harmony

and enhance both current and future potential to meet human needs and aspirations; all this means that human development must be done in ways compatible with ecological processes that support the work of the biosphere."[27]

Since then, publications on the concept of sustainability have multiplied and the subject of "renewable energy" has been discussed, including the problems that arise with its detection, exploitation and use. Probably the most systematic and concrete proposal has been made by the International Energy Agency (IEA).[28] A scenario called "450" is being built. The IEA's plan, expressed in the World Energy Outlook 2009 and entitled "the 450 Scenario", proposes an ambitious timetable of action which sets limits to the long-term concentration of greenhouse gases in the atmosphere of 450 parts per million of carbon-dioxide equivalent. This goal will also limit the global temperature rise to around 2°C above pre-industrial levels by the year 2030, which is considered sustainable. In fact, this goal becomes a paramount objective capable of keeping climate change under control.

- The 450 Scenario depicts a situation where, by 2030, energy efficiency will have induced over half the reduction in greenhouse gas emissions. In addition, by that time, low-carbon energy technologies will generate 60% of global electricity: renewable technologies (37%); nuclear (18%) and energy plants fitted with carbon capture and storage technology (5%). Finally, by 2030 car sales will have shifted dramatically, with hybrids, plug-in hybrids, and electric vehicles reaching nearly 60% of car sales (currently these vehicles represent 1% of car sales). The IEA estimates that this alternative will need an incremental investment of $10.5 trillion by 2030.

Nevertheless, research carried out by IEA and presented in the 2009 report, plus work done by the JELARE Project and other sources, shows many obstacles to reaching these objectives beyond advisable reductions, which puts us in a position to solve problems regarding this paradigm.[29] A synthesis of these problems is related to institutional aspects, which are the following:

- With just a few exceptions, efforts to promote renewable energies started in 2007, which is when oil prices reached unprecedented levels. Therefore, any assessment would have to be made in longer periods and even then, if the tendency continues, according to the IEA's primary demand projections fossil fuels would represent 80% of energy consumed and oil would reach 34% of the total world demand by 2030.
- The above would happen if current energy-use patterns continue. However, if a deliberate effort is made to reduce the use of fossil fuels and strive to attain sustainable climate change goals, extraordinary engagement would be required.

- A quick introduction of these efforts depends to a great extent on governmental support and a sound regulatory framework, given that renewable energy production costs are generally not competitive when compared with other sources of conventional energy at present.[30] This sole fact creates barriers that would take years to overcome. One industrialist participating in a focus group at the Universidad Galileo of Guatemala said:

  "The administration does not do its job right... it only expresses its will to promote renewable energies and never does, it's all talk" ... "the law is not done well and definitely favours fossil fuels" ... "I have a project of changing the regulation ... the law covers all, but the regulation only covers free projects. If you want a solar heater in your house, you have to pay taxes; however, big projects do have free access" ... "The government is good if it lets us work, if it does not block us ... for example, making revolving funds available for medium and small projects that, in general, do not have access to the big sources like the IADB or BCIE (the Inter-American Development Bank and the Central American Bank for Economic Integration) ..."[31]

- The gap between environment and renewable energy tends to be permanent and volatile. In many cases and countries, there is often confrontation between industrialists and investors making "clean energy" proposals, and the cause of environmental preservation, the latter sponsored by activists and national as well as international non-profit and non-governmental organisations. Often they involve the communities where natural resources are found, especially around hydroelectric power stations. Activists encourage communities, and investment is paralysed. Frequently the situation ends in violence. In general, ministries in charge of energy and mines and environment ministries work in different directions.
- In general, the common pattern is a lack of internal and external coordination within each sector (including the university sector), plus poor relations between the universities, the public and private sectors. This situation leads to the following relevant consequences, among others:
  - Little acceptance of external help.
  - Absence of programmes or volatility of existing programmes where coordination is needed between the government, the private sector, NGOs and universities.
  - Little synergy among current programmes inside universities and in different higher education institutions. This deficit might be extended to the private, public sector and non-governmental organisations.
  - Low investment in renewable energies due to lack of legal certainty and an unattractive environment for entrepreneurs.
  - Difference in the number of qualified personnel on these subjects and the demand of the market.

- With the exception of the more developed countries, renewable energy research and the consequent generation of patents, amongst other things, are virtually absent in the rest of the world. This leads to higher-cost technology for developing countries.

1. Renewable Energy and Universities

The postulates presented in the 450 Scenario focus on new sources of energy as an inherent part of its work. To further analyse the institutional problems that have been identified, universities become key in putting its agenda forward in this scenario. Its impact could be manifold. The universities might be able to play a role that could be described as follows:[32]

- There is a permanent demand for qualified personnel in renewable energy. The great majority of findings in the countries involved in the JELARE Project showed that people working in the renewable energy market, besides expressing the difficulty of finding qualified people to employ, considered that higher education institutions were failing to understand what they really needed.

- There is a need for a parallel transformation in the approach of faculties that have traditionally had renewable energy as part of their curriculum, such as in engineering and chemistry. It is necessary to broaden the spectrum of education in disciplines such as information technology, public policy, management, and social sciences such as economics, sociology, political sciences among others.

- To a great extent, the pursuit and acceleration of goals within the 450 Scenario depend on technological innovations emerging from research activities, which is virtually impossible in developing countries. Thus, emphasis on research is warranted.

- At the same time, it is necessary to identify and establish technology transfer mechanisms such as on-line educational programmes that allow the rapid dissemination of knowledge and, at the same time, might encourage research and innovations within the academic activity.

- Another aspect, which is not fully understood but that could mean a bridge between different sectors and increase coordination and synergy between different programmes, would be agreements and joint programmes between universities, certain private sector corporations, non-governmental organisations and institutions in the public sector, where theses and students' contributions could help. This would work together with internships that allow stu-

dents a gradual transition into the labour market and the operational objectives of these institutions.
- A less familiar aspect is the possibility of outsourcing specific tasks where there is a paralysis due to tensions between environmental and clean energy paradigms. That would mean appealing to universities as entities with pre-existing research capacities and conflict resolution techniques, in order to mediate between the opposing parties and thus provide an objective view on the existing differences that could accelerate investments in controversial areas.

IV. Conclusions and Recommendations

1. Fifty years of development have been analysed and three phases have been distinguished during the period 1960-2010.
2. Three visible phases have been detected: the "optimistic" (1960s-1970s), the "pessimistic" (1970s-mid-1980s), and the "realistic" (mid-1980s until the present day). Each of them represents an approximate period of time around the dates we have outlined.
3. The correlation between high oil prices and the promotion of renewable energy seems present at all times, but this trend currently appears permanent, given the uncertainty regarding the shortage of oil and the globalisation of environmental problems as a consequence of the use of fossil fuels.
4. The context of each development phase has also corresponded to the main policies that have attempted to counter both developmental and environmental problems. In the optimistic phase, the spirit of the time looked forward to future abundance, and believed in the persistent application of technology and human effort over time. In the pessimistic phase, the mantra was "putting the house in order", "setting growth boundaries, making "adjustments", reducing consumption, etc. Today, there is a call to pragmatism and the synthesis of previous phases. In this context the concept of "sustainable development" is thrown up, which aims to make striving for a better standard of living compatible with the preservation of the environment.
5. Nevertheless, the operative carriers of these ideas still encourage the arguments that made some paradigms valid during the periods examined.
6. Considering this situation, it is enlightening to go through the main paradigms that informed the different phases, which have been updated over time through different, yet related, intellectual developments.

7. Connections have been established between the promoters of modernisation theories and, more recently, the neoliberals, with an optimistic view of the development process that had its primacy in the sixties.
8. Some similarities have been pointed out between the theories that proclaimed the harmful effects of dependency in the development of countries on the periphery, and global theories that emphasised the extent to which we all depend on global vehicles of change (multinationals, investment, markets, etc.).
9. The above theory is reinforced by the work of the Club of Rome, which carried out, for the first time, a systematic analysis of the harmful consequences of all development efforts for the planet, from an environmental perspective.
10. Also, a correlation between these ideas and the intellectual "climate" of the time was noted: the modernisers belonged more to the optimistic stage, the "dependentists" and the followers of the Club of Rome to the pessimistic stage, and the sustainable development followers to the current realistic stage.
11. The most pragmatic aspect of the work that currently influences the "Sustainable Development" paradigm aims to broaden consensus around efforts to reconcile development and environment, which prevents many parties from participating with "maximalist" visions corresponding more to prior development stages when certain paradigms prevailed largely unchallenged.
12. The most advanced programme for the future has been presented by the International Energy Agency in Scenario 450, whose theoretical reflections and suggested alternatives put the "burden of the proof" on global decision-makers.
13. The difficulties of establishing institutional agreements to facilitate the 450 Scenario proposals were therefore advanced, highlighting that this programme is a sustainable and realistic goal whilst being attainable by the year 2030.
14. Inter-institutional coordination, both internal and external, is needed, aiming in the former instance to achieve inherent synergies of programmes and actions towards different programmes inside institutions, and, in the latter, to maximise relations between the private, public and non-governmental sectors, in order to attain the targets specified in Scenario 450.
15. Inter-institutional inefficiency weakens national capacity for securing external financing, especially important for research, to multiply and disseminate these objectives.
16. The role of universities in this context should place the emphasis on research, curriculum modernisation to encompass multidisciplinary approaches, identification of technology transfer mechanisms and internship agreements, and research contributions, considering close collaborations with other entities in the private, public and non-governmental sectors.
17. In this agenda it is worthwhile to note the need for convergence between "clean energies" promoters and those who proclaim their zeal for environ-

mental preservation in order to reach compromises in sustainable development-oriented concepts.
18. Irreconcilable positions are often assumed by these two groups, often paralysing pertinent action towards these goals, which must be made compatible with each other.
19. Universities have been shown here as institutions that, far from aligning themselves with any polarised attitude, might contribute to the enhancement of the 450 Scenario with their technical and scientific capabilities, providing the objective judgements and agreements that could allow advances in this field, through extensive research and suggestions for conflict resolution.

References

[1] Antecedents about this approach might be found in Nelson Amaro, "Contraste entre los Compromisos de las Cumbres Sociales y Paises Selectos" (A contrast between Social Summits Commitments and Select Countries), Seminario Sub-Regional de Capacitación. Los Acuerdos de la Cumbre Social. Implementación y Seguimiento, Post Ginebra 2000. Informe de Actividades (Sub-Regional Training Seminar. Social Summit Agreements. Implementation and Follow-up, Post-Geneve 2000. Activity report), United Nations Department of Economic and Social Affairs – UN DESA/Universidad del Valle de Guatemala, Guatemala City, Guatemala, November 30 to December 8, 2000, 33-47. See also Nelson Amaro, "Paradigmas del Desarrollo, Participación Ciudadana y Desarrollo Sostenible" (Paradigms of Development, Citizen Participation and Sustainable Development). Sustainable Development approach. Germán Rodríguez Arana, et al. Guatemala: FLACSO, 1999, 37-62.

[2] See G. Ritzer (2005, 1999). Teoría Sociológica Clásica. McGraw Hill, Mexico.

[3] Declaration on Social Progress and Development pproclaimed by General Assembly resolution 2542 (XXIV) of 11 December 1969.

[4] D. Suárez. "Historia del Petróleo" (History of Oil). Available at: http://www.monografias.com/trabajos72/historia-petroleo/historia-petroleo.shtml.

[5] K. Boulding (1966). The Economics of the Coming Spaceship Earth. Environmental Quality in a Growing Economy, Essays from the Sixth RFF Forum. Henry Jarrett, ed. Baltimore, Md.: Johns Hopkins Press, pp. 3-14.

[6] See Pedro-Pablo Kuczynski and John Williamson (2003). After the Washington Consensus: Restarting Growth and Reform in Latin America, Washington DC, Peterson Institute.

[7] R. Mostajo (2000). "Gasto Social y Distribución del Ingreso: Caracterización e Impacto Redistributivo en Países Seleccionados de América Latina y el Caribe" (Social Expenditure and Income Distribution: Description and redistributive impact in Selected Countries in Latin America and the Caribbean"), Economic Commission for Latin America, Series of Economic Remorms 69 LCL. 1376 (May 2000), p. 7.

[8] D. Sandalow (2006). "Ethanol: Lessons from Brazil". Seattle, WA: University of Washington College of Environment School of Forest Resources, pp. 1-2. Available: http://www.cfr.washington.edu/classes.pse.104/Assignments/Quizzes/bioethanolbrazil.pdf.

[9] T. Kuhn (1996). The Structure of Scientific Revolutions. Third Edition. University of Chicago Press, Chicago.

[10] See for example, Sanjay Reddy, Social Funds in Developing Countries, UNICEF STAFF WORKING PAPERS Evaluation, Policy and Planning Series no. EPP-EVL-98-002.

[11] These are summarised in the "Washington Consensus" stated in 1989: fiscal discipline, public expenditure priorities, tax reform, financial liberalisation, exchange rates, commerce liberalisation, direct foreign investment, privatisation, de-regulation and property rights. See John Williamson, "Revisión del Consenso de Washington" (Washington Consensus Review), "El Desarrollo Económico y Social en los umbrales del Siglo XXI" (Social and Economical Development at the threshold of the 21st century), Louis Emmerij and Jose Nuñes del Arce, Comps. Wash. DC: IADB, 1998. Actually, it is intellectualised in 1989 in Washington, but it had been stated in Bela Balassa, Gerardo M. Bueno, Pedro Pablo Kiczinsky and Mario Henrique Simonsen, Toward Renewed Economic Growth in Latin America. Mexico, Wash. DC: El Colegio de Mexico-Institute for International Economics-Fundaçao Getulio Vargas, 1986, and that circulated in 1986.

[12] E. Lora (2000). "What Makes Reforms Like it? Timing and Sequencing of Structural Reforms in Latin América". Inter-American Development Bank Interamericano de Desarrollo (VID), Research Department, Working Paper #424, June 2000.

[13] J. Langmore, "Social Development and the International Financial Systems", "Hacia un Sistema Financiero Estable y Predecible y su Vinculación con el Desarrollo Social" (Towards a Stable and Predictable Financial System and its Connection with Social Development), Series of Joint Subjects 8 (enero 2000) , Santiago de Chile: High profile meeting organised by the

Mexican Secretary of Foreign Affairs, with the support of the Economic Commission for Latin America and the Caribbean (CEPAL, for its initials in Spanish), Mexico DF, 49.

[14] C. Mandill (2010). Executive Director of the International Energy Agency, at the beginning of the decade in 2010, says: "Oil resources are ample, but more reserves must be identified to meet growing global demand to 2030 and beyond." Claude Mandill (2003). "The Oil Market: Conditions for a Stable and Sustainable Future". Middle East Petroleum and Gas Conference, Dubai, 7-9 September 2003, 2. Research by the IEA states that "The oil price in real terms is assumed to rebound from around $60 per barrel in 2009 with the economic recovery, reaching $100 by 2020 & $115 per barrel by 2030 in Reference Scenario". See Novu Tanaka (2010), current Executive Director of IEA, "Sustainable Energy and the Market", IEA/IEEJ Forum on Global Oil Market Challenges, February 26, 2010.

[15] K.J. Maynard (1936). The General Theory of Employment, Interest and Money, London: Macmillan (reprinted 2007), Book 6, Ch. 24 "Concluding Notes", p. 383.

[16] See T. Kuhn (1996). The Structure of Scientific Revolutions. Third Edition (Chicago: University of Chicago Press).

[17] Besides Rostow, quoted after, some other authors can be quoted. Examples of their work are: Talcott Parsons (1951). The Social System (New York: Free Press), pp. 45-67; also his "Pattern Variables Revisited," American Sociological Review, vol. 25 (1960), pp. 467-483, and "Some Considerations on the Theory of Social Change," Rural Sociology, vol. 26 (1961), pp. 219-239. Also Everett Hagen (1962). On the Theory of Social Change: How Economic Growth Begins, Homewood, Ill.: Dorsey Press; Daniel Lerner (1958). The Passing of Traditional: Modernizing the Middle East. New York, Free Press. Gino Germani (1974). Política y Sociedad en una Época de Transición. De la sociedad tradicional a la sociedad de masas. (Politics and Society in a Time of Transition. From a traditional society to a mass society). Buenos Aires: Paidos. And David Mclelland (1961). The Achieving Society. New York: Free Press.

[18] F. Fukuyama (1992). "El Fin de la Historia y el Último Hombre" (The End of History and The Last Man), Editorial Planeta, Buenos Aires, Argentina. Also Mario Vargas Llosa (1994). "América Latina y la Opción Liberal" (Latin America and the Liberal Option). INCAE, vol. VII, no. 2, Costa Rica; Lawrence E. Harrison (2006). The Central Liberal Truth. How politics can change a culture and save it from itself. New York: Oxford, and very influential at the time, Carlos Rangel (1976/2005). "Del Buen Salvaje al Buen

Revolucionario" (From good savage to good revolutionary). Caracas: Monte Ávila Editores. 1976 and then re-edited by Criteria, 2005.

[19] W.W. Rostow (1961). Las Etapas del Crecimiento Económico (The Stages of Economic Growth), (México: Fund for the Economical Culture) was probably the most influential book. Beyond its intellectual influence, Rostow held decisive positions when appointed National Security Adviser in John Kennedy's and Lyndon Johnson's administrations (1961-1969).

[20] The most influential author was Fernando Enrique Cardoso (1983), who later became President of Brazil. See "Dependency and Development in Latin America", Sociology of Developing Societies, Various Authors, London: McMillan Press Ltd.

[21] D.H. Meadow, D.L. Meadows, J. Randersf y W.W. Behrens III (1972). Los Límites del Crecimiento (Limits of Growth). Mexico: FCE, summary of these reflections.

[22] Immanuel Wallerstein (1976). The Modern World-System: Capitalist Agriculture and the Origins of the European World-Economy in the Sixteenth Century. New York: Academic Press; Giovanni Arrighi (2005). Rough Road to Empire. In F. Tabak (ed.), Allies as Rivals: The U.S., Europe, and Japan in a Changing World-System. Boulder, Colorado: Paradigm Press; André Gunder Frank (1991). El Desarrollo del Subdesarrollo. Un ensayo autobiográfico. Madrid: IEPALA.

[23] See for example José Luis de Imaz (1974). Adiós a la Teoría de la Dependencia, Estudios. Internacionales, vol. VII, no. 28, octubre de 1974.

[24] The often quoted criticism is in Robert Golub and Joe Townsend (1977). "Malthus, Multinationals and the Club of Rome," Social Studies of Science, vol. 7: 201-222.

[25] David R. Mares (2009). The Cardoso-Lula Paradigm for Growth and Energy Security, James A. Baker III, Institute for Public Policy, February 26, 2009.

[26] See F. Birol (2010). World Energy Outlook. Global Strategic Challenges. Available at: http://www.iaee.org/documents/washington/Fatih_Birol.pdf it was accessed on August 18, 2010.

[27] See Report of the World Commission on Environment and Development: Our Common Future. Transmitted to the General Assembly as an Annex to document A/42/427 – Development and International Co-operation: Environment.

[28] The International Energy Agency (IEA) is an intergovernmental organisation established in the framework of the Organisation for Economic Co-operation and Development (OECD in Paris) which acts as energy policy advisor to 28 member countries in their effort to ensure reliable, affordable

and clean energy for their citizens. Founded during the oil crisis of 1973-74, the IEA's initial role was to co-ordinate measures in times of oil supply emergencies. As energy markets have changed, so has the IEA. Its mandate has broadened to incorporate the "Three E's" of balanced energy policy making: energy security, economic development and environmental protection. Current work focuses on climate change policies, market reform, energy technology collaboration and outreach to the rest of the world, especially major consumers and producers of energy like China, India, Russia and the OPEC countries. The most recent meeting of the Governing Board of IEA member countries at Ministerial level was held on 14-15 October 2009 in Paris. With a staff of around 220, mainly energy experts and statisticians from its 28 member countries, the IEA conducts a broad programme of energy research, data compilation, publications and public dissemination of the latest energy policy analysis and recommendations on good practices. Available at: http://www.iea.org/about/index.asp.

[29] José Baltazar Salguerinho Osório de Andrade Guerra and Youssef Ahmad Youssef, organisers (2010). Renewable Energy Market Needs, a perspective from Europe and Latin America, Palohça, Ed. Unisul.

[30] See "Introducción, Monográfico, Energías renovables: presente y futuro (Introduction, monographic, renewable energies: present and future), Nota d'economía, Revista de economía catalana y de sector público (Catalonian economy and public sector magazine), 95-96 (1st Four-month period 2010) 5.

[31] See Cyrano Ruiz Cabarrús, Nelson Amaro, Robert Guzmán, Lourdes Socarrás y Ericka Tuquer (2009). Estudio sobre Energía Renovable y Mercado Laboral entre Universidades, Sector Público y Privado en Guatemala (Guatemala: JELARE-Universidad Galileo) 77.

[32] Many of these ideas, can be found in the Supervision Technical Team of the JELARE-Guatemala Project, Strategic Plan 2010-2012. Renewable Energy Capacity Building, Universidad Galileo, 2009.

Acronyms

BCIE	Banco Centroamericano de Integración Económica (Central American Bank for Economic Integration)
IADB	Inter American Development Bank
CO_2	Carbon Dioxide
COCODES	Consejos Comunitarios de Desarrollo (Community Development Councils)
USA	United States of America

IEA	International Energy Agency
IVA	Impuesto al Valor Agregado (Value Added Tax)
JELARE	Joint European-Latin American Universities Renewable Energy Project
ILO	International Labour Organization
WHO	World Health Organization
NGO	Non-Governmental Organization
UN	United Nations Organization
OPEC	Organization of Petroleum Exporting Countries
UNEP	United Nations Environment Programme
UNESCO	United Nations Educational, Scientific and Cultural Organization
UNFPA	United Nations Population Fund
USAID	United States Agency for International Development

"E-Learning: Sustainability, Environment and Renewable Energy in Latin America: a Multinational Training Pilot Module at Postgraduate Level"

N. Amaro, F. Buch, and
J.B. Salgueirinho Osório de Andrade Guerra[1]

Abstract

A pilot module will be implemented by four of the JELARE project's partners: Bolivia, Brazil, Guatemala and Latvia. Research was carried out by them in their own countries, where a scarcity of multidisciplinary programmes was detected at postgraduate level. The common characteristic was the need to modernize the curriculum by introducing a more diverse outlook. The definitive student profile should aim to provide skills useful to the private and public, non-governmental and academic sectors. The pensum will consist of 13 courses, identified on the basis of an analysis of the competencies needed, contained within three modules: sustainability, environment and renewable energy (with courses such as Sociology of Development and Global Challenges, Environment and Sustainable Development, Energy Matrix Planning, Energy Economics, Policies and Regulations on Energy and Environment as well as courses mainly devoted to renewable energy and its management). The postgraduate programme is to be implemented over four trimesters. Online teaching methods will be introduced high-

---

[1] N. Amaro is at Universidad Galileo in Guatemala, 7ª. Avenida, calle Dr. Eduardo Suger Cofiño, Zona 10, Ciudad Guatemala, Guatemala, CA. (Email: nelsonamaro@galileo.edu), F. Buch is at Universidad Católica Boliviana, (email: fbuch@ucb.edu.bo) and José Baltazar Salgueirinho Osório de Andrade Guerra is at Universidade do Sul de Santa Catarina (UNISUL), (email: Baltazar.Guerra@unisul.br). All of them are directors of universities involved in the Joint European-Latin American Universities Renewable Energy Project together with the University of Applied Sciences in Hamburg, Germany, and the University of Chile. The authors wish to acknowledge the contributions made by the Alpha III Programme of the European Union for its continuous support and Engs. Miguel Morales and Lourdes Socarrás (Guatemala) and Willy Tenorio (Bolivia) as well as Aleksejs Zorins, Director of the Latvian JELARE Project. All of them actively participated in the discussions held at the meeting we all attended in Florianópolis, Brazil, where the main components of this pilot module were defined. In addition, Robert Guzmán (Guatemala) helped with the first preliminary draft.

lighting self-study, cooperation and tutorial guidelines. The scheme will launch activities by 1 July 2011; the institutional framework for implementation is currently being worked out among all interested partners.

I. Background

In the international context, there has been a boost in the relevance of environmental and energy topics in academia, business and international governance. Rising awareness of the consequences of climate change, the green energy revolution and the increasing scarcity of water – the interrelation between economic growth and the environment and the challenge of achieving sustainable development that allows industrialised nations to follow their growth path and developing nations to reduce poverty and catch up with the developed world have gained priority in the political agendas of most countries of the world community and have contributed to the creation of a booming technology market.

Within the JELARE project, surveys were conducted on renewable energy and the labour market among universities as well as in firms from both the public and private sectors. The conclusions suggested a need to generate teaching and learning pilot modules to strengthen and deepen the academic programmes of the partner universities. One of the decisions that achieved a complete consensus was the implementation of an Integral Postgraduate Degree in Renewable Energies (RE), in an e-learning mode.

All of the partner countries involved in this pilot module 1, as it is called, namely Bolivia, Brazil, Guatemala and Latvia, are endowed with abundant natural resources and biodiversity, especially renewable energy sources. For this reason, it is necessary to reinforce these sectors with specialized human resources. However, a deficit has been detected in the areas of energy and environment in all of the said countries. If this gap between renewable energy and environment can be closed, it will allow capacities within the private sector to be developed and strengthen research activities in higher education institutions as well as in governmental bodies and international cooperation agencies. These countries lack consolidated educational programmes within this field at postgraduate level. There is also an information gap due to the lack of higher education institutions exclusively dedicated to research, and of programmes able to produce the required knowledge for the definition of public policies, development strategies, technological adaptations and innovations within this field.

In general, many efforts related to the environment appear in local media as confrontations between communities and hydropower plant projects. Frequently, these investments are regarded as environmental pollutants without their character as alternative or "clean energies" which are a response to the use of fossil fuels being analysed in depth. On the other hand, after examining the course requirements of the study programmes comprising environmental, sustainability, climate change and renewable energy topics, among others, in different universities, it can be seen that there is a need to generate multidisciplinary content to overcome the gaps between what is purely environmental and what is strictly mentioned as the electricity or energy sector. In general, RE tends to be concentrated in engineering programmes and frequently is not associated with environmental studies or other disciplines of human sciences or information technologies.

These observations may be extended to general multidisciplinary knowledge such as demographics, economics and social, political and cultural sciences as they relate to development processes which tend to influence the energy mix. Also, the influence of public policies and the role of the public sector as a coordinator between the private, public and civil society activities in the promotion of alternative energies tend to be forgotten.

The joint postgraduate programme entitled "Sustainability, Environment and Renewable Energy", which will be offered in cooperation between the said partner countries of the JELARE project, may contribute to the formation of qualified human resources in an area which is relevant for socio-economic development, ranging from the reduction of dependence on imported liquid fossil fuels to the fostering of rural development, the creation of local jobs and the diversification of the economy. The programme will also build human resource capacities with the ability to generate knowledge and information at the academic level which will eventually improve political decision-making processes as well as technological adaptations and innovations.

As the postgraduate programme is executed successfully in its first version, there is a possibility of extending it to a two-year Masters degree programme that might combine specific competencies within disciplines such as economics, business administration, sociology, law and engineering, with the inclusion of the environmental topic as a transversal concern of all modules. These educational efforts are designed in line with the model of education aimed at the development of competencies, i.e. skills that go beyond the acquisition of subject-related knowledge. The required competencies can be divided into three main areas: projects/economics, technology/engineering and environment/sustainable development. It is this multidisciplinary approach that will enable graduates to meet the challenges of the public sector as well as the demand of NGOs, international cooperation and private businesses.

A. What might a professional who has graduated from the Postgraduate degree in Sustainability, Environment and Renewable Energy asked to perform?

These observations aim to highlight the following dimensions demanded by the market:

Research

The graduate professional should have acquired capacities in research and analysis.

Diagnostics in the area of energy generation

Therefore, the postgraduate degree should have as its objective the formation of professionals specialized in resource evaluation, design, technical and economic viability analysis, optimization and management of renewable energy technologies.

Environmental impact evaluation

Graduates should understand and have knowledge of how to apply the fundamentals of environmental impact evaluation, the general concepts that rule this field, and the management of its main tools.

Preparation and evaluation of public policies

The professional should be familiar with the main concepts relating to public policies, the relationship with the legal system of the host countries and the global covenants regarding environmental protection. Furthermore, professionals should be aware of the fiscal and legal instruments and other norms applied in the partner countries and elsewhere.

Preparation of strategic plans in the area of renewable energy

This is an ability to generate strategic plans that should encompass the integral aspects of the renewable energy subject, where socio-economic, environmental, legal and other disciplines are more than necessary elements for an adequate planning that comprises elements such as energy planning, energy economics, environment, etc., aiming to achieve a desirable energy matrix within a determined period of time.

Project management, including renewable energy firms

Such a scheme should provide professionals interested in the postgraduate degree with the managerial tools that will enable them to administer, plan, organize and manage projects and firms in the preparatory phase and pre-investment and investment process. Graduates should be capable of conducting programmes, projects, plans, etc. in the energy sector in an integral manner.

Updates regarding new technologies

Any professional should have information regarding new technologies that are being continually developed within the sector of the utilization of natural resources for energy, including those involving information technologies that give access to this knowledge.

II. Characteristics and profile

This section will examine the added value of the degree, the general and specific objectives, the professional profile and the requirements needed to access the degree as well as the procedures for enrolling in the career.

A. Added value of the degree

This degree aims to be an exchange of experiences in the field of e-learning between the following universities: UNISUL in Brazil; the Bolivian Catholic University; Galileo University of Guatemala; and Rezekne University of Latvia, which are part of the consortium of universities making up the Joint European-Latin American Universities Renewable Energy Project, JELARE, mostly financed by the European Union under the Alfa III Programme. Such exchanges, based on the experiences and developed technology of these higher education institutions, can strengthen the online teaching already being practised in many of these universities and extend this expertise to those that do not have this technology.

Likewise, an educational component will be generated that will provide knowledge to the people interested in it in an e-learning mode. Participants will be exposed to the curriculum of a postgraduate degree in Sustainability, Environment and Renewable Energy which might subsequently be extended to a Masters programme. On the other hand, there will be an analysis of the factors that influence the supply and demand of energy within industrial societies and developing countries which are eager to produce and consume energy that respects the requirements of a sound environment. The integral and interdisciplinary characteristics of the postgraduate degree may allow professionals to obtain a wider knowledge which will enable them to move between several positions in various fields of work. This characteristic makes the described degree additionally attractive.

B. General objective

The general objective of the e-learning pilot module entitled "Postgraduate Degree in Sustainability, Environment and Renewable Energy" is to increase the capabilities of the partner universities in virtual education and to implement a postgraduate multidisciplinary study programme relating to the environment and renewable energies in these universities and with other bodies and individuals which eventually might become partners.

C. Specific objectives

1. Develop a postgraduate programme within the field of sustainable development, environment and renewable energy.
2. Develop virtual educational material of high quality related to these topics.
3. Implement this postgraduate programme in "Sustainability, Environment and Renewable Energy" jointly in an e-learning mode among the Bolivian Catholic University, the Universidade do Sul de Santa Catarina of Brazil, the Galileo University of Guatemala and the Rezekne University of Latvia.
4. Conduct an evaluation of the first version of the postgraduate programme in order to improve it, ensure its sustainability and possibly extend it to a Masters degree programme.

D. Expected outcomes

At the end of the implementation of the proposal, the following products are expected:

1. A joint study programme consisting of an international postgraduate degree in "Sustainability, Environment and Renewable Energy", implemented by Universidade do Sul de Santa Catarina in Brazil, the Bolivian Catholic University and the Galileo University of Guatemala with some support from the Rezekne University.
2. 13 online study courses developed.
3. At least 30 graduates at postgraduate level, ten for each of the participating universities.

E. Professional profile and requirements to enrol

A graduate of the postgraduate degree in "Sustainability, Environment and Renewable Energy" will be capable of developing projects that support and promote renewable energy sources and the environment in an integral manner. Graduates will also be capable of generating policies that could contribute to the protection of natural resources. These capabilities will be assets that will enable graduates to perform as a consultant for agencies concerned with the environment, natural resources, renewable energy, climate change, etc. In this sense, the following characteristics are required from students in order for them to truly take advantage of the programme:

- Understanding of the need for a rational and efficient use of all types of energy, fossil or renewable, in order to achieve a more sustainable human development.
- Awareness of the current and future situation of the energy market in a regional and international context and the consequences of the limits, conflicts and impacts of fossil energy for the environment and sustainable development.
- Establishment of a clear perspective of the possibilities and economic viability of renewable energies, linking the multidisciplinary knowledge (social, instrumental and technological) acquired to the environment and sustainable development.
- Detection of environmental threats at national and global levels.
- The basic knowledge to develop a professional activity within the field of the installation, operation, management and maintenance of renewable energy systems, with a basic training regarding the different technologies of these systems.
- Knowledge of the normative and regulatory framework of renewable energy and the environment.
- Awareness of the criteria of energy savings and efficiency enabling him/her, in the exercise of his/her professional duties, to bring about the improvement of the existing energy installations based on the use of fossil fuels.
- Knowledge of the sources of information required remains up-to-date on a permanent and continuous basis as well as of specific tools for searching the relevant information. The objective is to create capabilities for professionals so that they can find the best responses to the problems they face, adapted to their own reality.
- Openness to integrate energy efficiency, renewable energies and energy management, from the perspective of sustainability and an environmental approach in an integral way that is capable of incorporating other fields of knowledge.

The aforementioned criteria will be the topic of a personalized interview with each student and will serve as a guide to identify the student's potential and capability to develop the competencies that the student wishes to learn.

F. Requirements for enrolment (regarding the enrolment procedures)

1. Graduation at a Bachelors or licenciatura's level in related areas.
2. Knowledge and experience in the study areas.
3. Willingness to learn about the required e-learning tools through a personalized interview.
4. Interest in the e-learning mode.

III. Generation of the curriculum based on the required competencies

The required competencies to enrol in the postgraduate degree in "Sustainability, Environment and Renewable Energy" are the result of a study conducted by many countries belonging to the European Union. This study was later extended to Latin American countries in order to fine-tune the competencies and confirm them.[2] The definition of competencies established for Europe and subsequently applied to Latin America, is as follows:

> "Competencies represent a dynamic combination of knowledge, comprehension, capabilities and abilities. To foster them is the purpose of educational programmes. Competencies are formed in several course units and are evaluated at different stages. They can be divided according to whether they are specific to an area of knowledge (field of study) or generic (common to different courses)."[3]

A test identifying competencies was undertaken by multiple careers in this extended study that included the majority of the universities of Latin America. The

---

2   See Pablo Beneitone (Argentina), César Esqueitoni (Ecuador), Julia González (Spain), Maida Marty Maletá (Cuba), Gabriela Sufi (Argentina) y Robert Wagenaar (The Netherlands), Eds., Reflections and perspectives of Higher Education in Latin America, Final report Tuning-América Latina, 2004-2007 (Spain: Universidad de Deusto-Universidad de Groeningen, Project financed by the Alfa Programme of the European Commission, 2007. This research reached (182) universities of almost all of Latin America. Its objective as stated by its text "is to identify shared competencies, that can be generated at any title and that are considered important for certain social groups." (page 15). Available in: http://tuning.unideusto.org/tuningal/index.php?option=content&task=view&id=217&Itemid=246.

3   Tuning Report, p. 37.

following have been selected for the postgraduate degree, considering only those that were closer to our goals:

A. Generic competencies

1. Capacity for abstraction, analysis and synthesis
2. Social responsibility and commitment to citizenship
3. Ability to use of information and communication technologies
4. Commitment to look after the environment
5. Commitment to socio-cultural environment

B. Specific competencies

6. Improve and innovate administrative processes using information and communication technologies for the processes which allow for its formulation and optimization.
7. Awareness of the responsibilities regarding the environment and the values of urban and architectural heritage as well as the capability of knowing and applying research methods to resolve creatively the demands of the human habitat, in different scales and complexities.
8. Ethical commitment regarding the discipline, manifesting a social consciousness of solidarity and justice, and respect for the environment.
9. Provide advice regarding the use of natural resources in the formulation of development policies, norms, plans and programmes, interacting in interdisciplinary and trans-disciplinary areas.
10. Development of professional activity within a framework of responsibility, legality, security and sustainability, when planning, executing, managing and supervising projects and services focused on the knowledge, use and exploitation of renewable natural resources.
11. Suggest solutions that might contribute to sustainable development, planning, research design and execution with regard to the topic in question.

In Appendix 2, there is a list of 11 characteristics that summarize the concept of competencies. They served the purpose of selecting the courses to be offered. The concept of competencies will be continuously used to ensure the excellence of the programme while the initial selected matrix of course might be improved over time. They will also serve as guidelines for evaluating the implementation of the postgraduate degree as a whole in order to determine later, in 2011, whether the pilot module might be extended to a Masters programme for the second year. Appendix 2 shows how the selected courses are adapted to the competencies concept.

The courses which finally form part of the postgraduate degree were previously assessed in each university with regard to the extent that these competencies had an impact on the courses selected. The installed capacity of each institution for delivering this learning product was also considered. The next step was a meeting of all participating universities to choose those courses from the total sample which would finally form part of the online postgraduate degree. To this end, a meeting was held in Florianópolis, Brazil. The outcome of this meeting was a definition of the programme's objectives, the selection of the courses, a division of labour among universities and a schedule of activities up to 1 July 2011, the date of the programme's launch.

IV. Description of the curriculum

After the selection of the courses that will form part of the degree, a description of each of them is warranted:

A. Sociology of Development and Global Challenges

This course anticipates that the student has had or will be open to an Introduction to Sociology or elements of General Sociology. The content will examine the perception of change by the classic authors, ranging from Comte, Spencer and Marx to Weber. It will venture into more recent middle-range theories that emerged during the Sixties with Lerner, Hagen, McLelland and others. It will examine the reflections of the Club of Rome and the Dependency Theories characteristic of the 70s and 80s. Afterwards, it will explain the theories of the global system and finish with the post-modernists and considerations regarding sustainable development which have been discussed since the 1990s and still are today. These lessons will establish links with the topics of renewable energies.

Global challenges will be illustrated with continuous references to the development patterns predominant in the least developed countries. In particular, the emergence of Asian countries in that context will be analysed and compared with the situation of Latin American countries. The focus will be on the role of foreign aid, commerce, foreign direct investment and their impact on social structures and political developments. Within this framework, the dimensions of demography, gender and environment and recent challenges will be integrated. Subsequently, the main theoretical and conceptual problems with regard to the sustainable development paradigm will be highlighted and their interrelation with renewable energy will be made evident.

B. Environment and Sustainable Development

Introduction to the Environment: The concept of environment, the systems of planet Earth, ecosystems, historic evolution of environmental concerns, global environment, the relationship between human beings and the environment. *Introduction to Sustainable Development:* The concept of sustainable development, the history of sustainable development from Rio de Janeiro to Cochabamba, Agenda 21, the Kyoto Protocol, environmental indicators, sustainable development in developed countries and developing countries. *Sustainable Development, Natural Resources and the Environment:* Poverty and environmental degradation, international trade, growth and environment, loss of biodiversity, climate change, mitigation and adaptation, the role of natural resources. *Sustainable Development and Renewable Energies*: The effect of energy consumption and environmental problems, renewable energies and sustainable development, future prospects for sustainable development.

C. Planning of the Energy Matrix

Introduction to energy policy. Knowing the bases for the development of policies that reinforce the sustainability of the energy sector. Elements for the design of an energy policy. Analysis of energy and integration policies. Energy planning, integrated plans for resources. Investment planning. Energy sustainability with emphasis on energy efficiency policies, the obstacles for the efficient use of energy. *Energy Planning Tool.* Introduction to the tools used for energy planning. Comparative analysis of energy planning models, case studies. Selection of energy planning tools. Use of programming models and detailed operation of *LEAP* (long-range energy alternatives planning system) whose main objective is to bring integrated and reliable support for the development of integrated energy planning studies.

D. Environmental Management and Impact Evaluation

Environmental management. The distribution of competencies within legislation, planning and management of the environment at national and international level. Basic knowledge regarding environmental legislation. Characteristics and principles. Management instruments. Environmental management in the company. Environmental responsibility. Administrative, civil and legal solutions. Access to environmental information. Current legal framework for the environment. ISO norms. Design of an environmental management system. The environmental audit as an instrument for the company's environmental management. Ecological marketing as an instrument of environmental management. Competence and awareness raising. Communication. Operational control. Preparation and emergency control. Verification. Follow-up and measurement. Evaluation of legal compliance. Non-conformity. Corrective and preventive action. Internal audit. Revision by the direction. Ecological differences of processes and products.

With regard to impact evaluation: Conceptual, legal and institutional framework. Introduction to environmental impact studies. Technical document of project analysis. Identification and evaluation of environmental impacts. Preliminary environmental impact study. Partial environmental impact study. Baseline study or socio-environmental diagnostics. Strategic environmental evaluation. Preventive and corrective measures. Surveillance plan and environmental control. Management procedures of environmental impact studies.

E. Policies and Regulations for Energy and the Environment

Principles of Energy Policy: Analysis of the different principles and criteria of energy policy, environmental objectives within energy policy, scenario analysis and energy policies. *Energy Intensity:* Analysis of energy intensity by sectors that demand it. *Regulation of Tariffs and Prices within Energy Markets:* Bases of regulation, roles of regulatory organisms, structural analysis of tariffs and prices. *Principles of Environmental Policy:* Analysis of the different principles and criteria of environmental policy, scenario analysis of environmental policies. *Instruments of Environmental Policy:* Moral persuasion, environmental norms, economic instruments (taxes, subsidies, emission trading).

F. Energy and Environmental Economics

Introduction to Energy Economics: General aspects of energy, types of energies, energy units. *Energy Trade and Environmental Services:* Conventional and renewable energy commercialization methods, forms of concentration and fixation of prices in different conventional and renewable energy markets, carbon markets and of environmental services. *Analysis of Energy Supply and Demand:* Technical structure of conventional and renewable energy sectors, their economic structure, peculiarities and environmental incidence. Sectors with intensive energy demands, energy costs according to products and processes, environmental impact of energy demand. *Introduction to Environmental Economics:* Externalities, public goods, the Coase theorem, optimum level of pollution. *Economic Valuation of Environmental Quality:* The value of the environment, environmental valuation methods. *Economic Development and Environmental Quality:* Economic growth models that incorporate energy and environmental restrictions, the energy-environment relationship.

G. 4.7 Solar Energy

Fundamentals of Solar Energy: Role of solar energy within the international energy mix. Energy savings and efficiency. Description of the sources of thermal and photovoltaic solar energy and the design, maintenance and operation of installations. Advantages and disadvantages. Environmental, social and economic impact of the technologies. *Solar Thermal Energy:* Solar collection system. The storage and accumulation sub-system. Performance. Description and design of solar thermal installations. Evaluation of the environmental impact of solar thermal energy. Perspectives and development of the legislation regarding solar thermal energy. *Photovoltaic Energy:* Applications of photovoltaic energy. Fundamentals of photovoltaic energy. Components of photovoltaic installation. Design and calculation of installations. Exploitation and maintenance of an installation. Environmental impact of photovoltaic energy.

H. Hydropower

Role of hydropower within the international energy mix. Energy savings and efficiency. Description and design of installations, maintenance and operation. Advantages and disadvantages. Environmental, social and economic impact. The role of hydroelectric energy. Electro-mechanic systems. Environmental impact. Legal and normative aspects. Criteria for the development of hydro power projects. Tools for preparing projects of hydropower stations. Feasibility study sample.

I. Biomass Energy

Role of biomass in the international energy mix. Energy savings and efficiency. Description of the different sources of biomass and the design, maintenance and operation of their installations. Advantages and disadvantages. Environmental, social and economic impact of each of them. Biomass classification. Biomass sources. Physical and chemical characteristics which define a fuel. Processes of conversion of biomass into energy. Energy applications of biomass. Advantages and disadvantages of the use of biomass. Legislation, incentives and fiscal measures.

J. Wind Energy

Role of wind energy within renewable energies in the international energy mix. Energy savings and efficiency. Description of the different sources of renewable energies and the design, maintenance and exploitation of their installations. Advantages and disadvantages. Environmental, social and economic impact of each of them. Historical evaluation of the use of wind. Meteorological bases for wind energy. Use of wind. The wind potential. Wind generator: composition and function. The wind park. Off-grid wind power installations. Offshore wind energy. Wind energy and the environment. Stages of the development and management of a wind energy project. Legislation.

K. Energy Efficiency

Basic definitions: Energy sources: Primary/secondary. Renewable/non-renewable, *Energy systems:* Primary energy, production and conversion of sources in energy carriers, transport and distribution of energy carriers, net energy. Final use of the energy. Useful energy, supplied service, received benefit. *Flow of energy:* Unit operation, global energy performance. *Energy efficiency: General bases and measurements of EE:* Good operational practices; closed circuit of recycling; substitution of energies; modification and optimization of processes; product reformulation; technological improvement/substitution; *The Energy Diagnostic:* Unitary operations; process flows; focus on diagnostic; balance of energy; thermal energy; electric energy; identification of losses/inefficiencies; consumptions, emissions and specific costs; critical unitary operations; energy efficiency measures; technical – economic evaluation; *Efficiency of the productive processes. Application examples:* Considerations of EE within the energy mix; EE in a system of electricity distribution; Measures of EE in productive systems.

L. Renewable Energy Project Management

The students will be prepared as managers of renewable energy projects and firms, developing capabilities of conceptualizing and managing this type of projects within the current economic scenario. Economical and legal aspects which allow for the development of own business initiatives within the sector. Organization, planning and coordination of projects of diverse complexity through an ample study of experiences, techniques, tools and methodologies related to project management. Viability and design. Business opportunities, profitability and opportunities for financing. Legal procedures, permits and operations. Analysis of suppliers and products. Management tools: Integrated management of projects. Project planning management. Project cost management. Product quality management and energy efficiency. Project resource management. Project human resource management. Project communications management. Project risk management. Project acquisitions management. Analysis for the reduction of emissions.

V. Modules by Courses

The courses described above pertain to three areas which might become modules. Table 1 shows the relationship that helped to give the postgraduate programme its title.

Table 1: Postgraduate Modules

	SUSTAINABILITY		ENVIRONMENT		RENEWABLE ENERGIES
1	Sociology of Development and Global Challenges	4	Environmental Management and Impact Evaluation	7	Solar Energy
2	Environment and Sustainable Development	5	Policies and Regulations for Energy and the Environment	8	Hydro power
3	Planning of the Energy Mix	6	Energy and Environmental Economics	9	Biomass
				10	Wind Energy
				11	Energy Efficiency and Renewable Energy
				12	Renewable Energy Project Management
13	Research Methodology				

Table 2 shows the distribution of courses by trimesters.

Table 2: Distribution of courses by module and trimesters

Modules and courses	Trimesters				Total courses
	1st	2nd	3rd	4th	
Sustainability	Sociology of Development and Global Challenges			Planning of the Energy Mix	2
Sustainability	Environment and Sustainable Development				1
Environment		Environmental Management and Impact Evaluation	Policies and Regulations for Energy and the Environment		3
Environment		Energy and Environmental Economics			
Energy	Biomass	Solar Energy	Hydropower	Wind Energy	4
Energy			Energy Efficiency and Renewable Energy	Renewable Energy Project Management	2
Common Area		Research Methodology	Preparation of the final thesis	Preparation of the final thesis	1
Total	3	4	3	3	13

VI. Institutional Framework

The postgraduate degree is a joint effort and as such the work is shared. The following courses will be prepared by the partner universities according to Table 3 in the corresponding countries (the numbering corresponds to the course description stated above).

Table 3: Responsibilities of the participating partners as regards courses

Bolivia	2) + 5) + 6)	3) + 11) will be shared between Guatemala and Bolivia. See note)
Guatemala	1) + 4) + 12)	
Brazil	7) + 8) + 10) + 13)	
Latvia	9)	

Note: Courses 3 and 11 will be shared, with the main responsibility for 3 falling on Guatemala with the support of Bolivia and for 11 on Bolivia with the support of Guatemala.

VII. Periods and Academic Credits

Duration: 1 year divided into 4 trimesters of ten weeks each.
Frequency: See Table 4
Hours per week: See Table 4
Distribution of academic credits: See Table 4

The academic credit is a measurement of the students' working hours to achieve learning goals and allows studies completed in several institutions to be compared and approved. It is also an efficient instrument for achieving of curricular flexibility and planning of the study programme. The credits as well as the assigned hours are detailed in Table 4.

Table 4: Calendar per trimester and academic credits

Trimesters	Course	Academic Hours	Credits
I – Trimester Jul–Aug–Sept	Sociology of Development and Global Challenges	30	2
	Environment and Sustainable Development	30	2
	Biomass	30	2
II – Trimester Oct–Nov– Dec	Environmental Management and Impact Evaluation	30	2
	Energy and Environmental Economics	30	2
	Solar Energy	30	2
	Research Methods	30	2
III – Trimester Jan–Feb–Mar	Policies and Regulations for Energy and the Environment	30	2
	Hydropower	30	2
	Energy Efficiency and Renewable Energy	30	2
IV – Trimester Apr–May–Jun	Planning of the Energy Mix	30	2
	Wind Energy	30	2
	Renewable Energy Project Management	30	2
	Thesis at the end of the prior courses	0	0
TOTAL		390	26

Note: For lectures, reports and other curricular activities, an average of four additional hours is estimated for each course, which will require each country to provide 442 hours of teaching and personal study in a uniform manner.

VIII. Academic Methodology

A. Research

This component will be present in the application of the online methodology, it being expected that once the curriculum is finished, the student will complete his/her effort with a paper in an area of his/her interest. Likewise, all of the professors will place emphasis on the application of research methodologies, which in themselves are an important part of the curriculum content.

B. Participation: Essential Characteristics of the Programme

The expert or specialist who works as a professor on the course is primarily a facilitator of student's self-study and research activity; the main feature of this profile is not face-to-face presentation. Nevertheless, the online mode must also allow for interaction between students and professors and among students.

C. The Balance Between Theory and Practice

The direction of the degree will ensure that training and activities maintain a balance between theory and practice. Managerial aspects, practical knowledge and systematic actions in real-life situations will be part of the courses.

D. Evaluation

In line with the dynamic, participative and balanced character that the programme wishes to establish, the evaluation exercise will emphasize academic excellence, comprehension, efficiency, feasibility and viability.

IX. E-Learning Methodology

The aforementioned principles must be seen in light of the e-learning mode that will be applied.

A. Learning Mode

The postgraduate degree will be offered entirely in e-learning mode due to two primary motivations:

1. Being able to have a group of excellent experts with multidisciplinary approaches focused on a very specialized topic like renewable energy will encourage the construction of different scenarios.
2. Allowing a wider application, taking into account that, with e-learning, several countries of the world can be reached, allowing for interaction between different peoples and cultures.

The courses offered within the e-learning mode are educational concepts that integrate technological, didactic and administrative support to extend and transfer the contents of any subject of knowledge. These types of courses are based on the application of new Information and Communication Technologies (ICTs) which allow for learning without limitations as to place, time, occupation or age of the students.

B. Principles of the Mode

- Self-study: The course materials as well as the greater part of the practical activities are designed in a way that enables the student to advance at his/her own pace and assess his/her progress at any time.
- Teamwork: The student will not learn in an isolated manner; part of the knowledge will be constructed by the group thanks to the interaction with the rest of the course members.
- Tutorial support: The tutor will guide the group in the learning process, conducting an individual follow-up of their participation, efforts and results during the course.

C. Characteristics of the Model

- The students' participation is not passive – they become the protagonists of the teaching/learning process.
- It is important how the students learn and not how the teachers teach.
- The tutor plays a guiding role.
- It is not suited to all educational levels because it requires much discipline, maturity and commitment.
- The learning must guide the student towards reality.

- More responsibility from the student in the learning process.
- Flexibility in time management. Nevertheless, this does not imply an absence of deadlines for learning activities.

D. Structure, Characteristics and Resources

All courses have been developed by professionals in the subject. Each of the syllabuses is adapted to practice in a way which ensures that they end up being interesting, enjoyable and practical. The common structure is as follows: introduction, contents, activities (case studies), annexes, bibliography and glossary.

Furthermore, each topic comes in the section on Activities with questions that allow the student to fine-tune his/her knowledge and measure his/her rhythm of study. There are also exercises that allow the student's skills to be evaluated. The team of tutors, specialists in the different areas of study, will pay attention to students through email, forums or chats and, if necessary, with a synchronous meeting (video conference).

Learning activities

- Forums, homework, exercises, fieldwork, research, case study
  Use of resources to make the readign of what is below.
- Videos, presentations, audio, animations

Student rate per e-moderators

- One for 20 or 30 students

Platform

There are services provided, such as:
- *Communication services:* discussion groups, forums or news, chat or interactive talks, email, working groups, etc.
- *Evaluation services:* tests, questionnaires, auto evaluations, report cards, monitoring tools, wiki.
- *Information services:* glossaries, dictionaries, etc.

Online academic periods

There is an estimate of an average of 390 tutorial hours plus other 52 hours dedicated to reading, studying, discussions, reports and essay elaboration, field works, elaboration and drafting of a final research project. The total will be 442 hours.

E. A Suggestion for the Marketing Design

An important aspect to consider is the promotion of the postgraduate degree, and even more relevant is contemplating the prospective student. Therefore it is recommended that an induction document be created with the objective of showing the most relevant characteristics of the model. Here are some of the considerations that must be taken into account:

Table of contents

a) Welcome
b) About us
c) How to study?
d) What do you need?
e) Learning about the e-learning mode:
    e.1 Why?
    e.2 Where?
    e.3 When?
    e.4 How?
f) Advantages

F. Start of the Postgraduate Course

Regarding implementation, it is anticipated that the degree of development of the various virtual platforms in the participating universities will provide the opportunity for a technology transfer from one university to another on the base of a collaboration agreement. The agreement will include training, graphic design and instructional methodologies, etc.

Another critical point is the development of content based on a uniform model as well as their virtualization, which will be conducted by experts in the area. It is also necessary to fine-tune the costs and make them uniform for each responsible entity in the universities. A gross estimate of the programme's cost makes it a strategic period for return on investment in the design of contents, instructional advice, and the assembly of content incorporated into virtualization and graphic design (multimedia). The investment during the first year will be the greatest; it will become lower over time with use. This calls for the need to contemplate a feasibility study of the administrative and financial management of this modality.[4]

In the meeting at Florianopolis on 4–8 July 2010, a schedule of activities was developed for implementation in a joint and individual manner. The date for launching of the postgraduate degree in Sustainability, Environment and Renewable Energy was set at 1 July 2011.

Conclusions

This exercise has been reported to highlight the importance of a thorough preparatory approach whenever a new career modality is attempted. Several lessons could be learned from this outcome that might be useful for other similar experiences:

1. Usually, online e-learning is centralized. This characteristic makes planning and implementation easier, which has its advantages. Nevertheless, when this is done by a network of institutions, although introducing the complexities of coordination across institutions and countries, a division of labour among partners considerably reduces costs. On the other hand, strengths developed by different institution when put together, improve the quality of the courses and the overall design.
2. The international flavour that has been added to the design is an additional attraction for enrolment. The endorsement of universities from other countries is an aggregated value which enhances the promotion of e-learning arrangements.
3. Innovations in the content of the courses call for a multidisciplinary effort that breaks down the frontiers between different disciplines. This outcome is proof that this is possible and that collaboration between engineers, economists, environmentalists, social scientists and political scientists may produce a fruitful outcome.
4. The above description of an e-learning postgraduate programme on Sustainability, Environment and Renewable Energy could have followed different methodologies in order to put together a specific course of study. The simplest option is to assemble a collection of courses without questioning further objectives such as to what kind of competencies these courses belong or how will the labour market receive a graduate with such qualifications, or without

---

4   To give an example regarding the costs of the programme, a similar programme in the Galileo University of Guatemala, a Masters in Telecommunications, has an approximate total cost for the student of US$728.0 per quarter, including US$130.0 per enrollment, 3 payments per month of US$196.0 for all the courses corresponding to the quarter and US$10.0 dollars for electronic services and ID. Per annum it would be approximately US$2912.

considering a permanent evaluation of the suitability of those courses with regard to the competencies that the modality aims to enhance. The design presented here has made a frontal incursion on these quality avenues that should be a permanent challenge for any similar exercise in any career.

5. Another lesson might be inferred for the future with regard to the external cooperation enabled by the European Commission. Sometimes, such assistance lasts as long as funded project continues, without considering sustainability plans beyond the life of the project. In this case, the external assistance made mutual exchanges possible across countries and institutions for a limited period, but took into account the fact that the initial help would eventually place responsibility for the postgraduate programme's full implementation on the institutions' shoulders in each university and country. Therefore, sustainability is ensured beyond the life of the project.

References

Pablo Beneitone (Argentina), César Esqueitoni (Ecuador), Julia González (Spain), Maida Marty Maletá (Cuba), Gabriela Sufi (Argentina) y Robert Wagenaar (Netherlands) (eds.) (2007). Reflections and perspectives of Superior Education in Latin America. Final Report-Tuning Project-Latin America, 2004-2007. Universidad de Deusto-Universidad de Groeningen, Project financed by the Alfa Programme of the European Commission.

Guy Le Boterf (2000). Competence Engineering Barcelona, Spain: EPISE, Training Club, Editions Period.

Laura Hersh Slganik, Dominique Simone Rychen, Urs Moser, and John Konstant (1999). Projects regarding competence within the Context of the OECD. Theoretical and Conceptual Base Analysis. Neuchatel, Switzerland, FS/BFS/UST-OCDE-ESSI. Available at:_http://www.scribd.com/doc/18765954/Proyectos-sobre-Competencias-en-el-Contexto-de-la-OCDE.

Appendix 1: Postgraduate examples in countries of Latin-America and the world

Mexico:
Masters in Environment and Renewable Energies, http://www.lumni.com.mx/articulos/index.php?consecutivo=526&se=54&ca=
Renewable Energies, http://maestria.emagister.com.mx/maestria_energias_renovables-cursos-797173.htm

Paraguay:
Masters in Energy for Sustainable Development, Renewable Energies and Energy Efficiency, http://estudios.universia.net/paraguay/estudio/uc-maestria-energia-desarrollo-sostenible-energia-renovables-eficiencia-energetica

Argentina:
Masters in Renewable Energies, http://www.universia.com.ar/contenidos/buscador_carreras/form_alf.php

The UNIVERSIA network has online programmes in the following countries of Latin America: Argentina, Bolivia, Brazil, Chile, Colombia, Costa Rica, Ecuador, El Salvador, Guatemala, Honduras, México, Nicaragua, Panama, Paraguay, Peru, Puerto Rico, Dominican Republic , Uruguay and Venezuela.

Examples of postgraduate degrees in European countries:

In *Spain* the following online degrees are examples of what can be achieved due to the simplicity of implementing degrees in an e-learning mode.
Postgraduate Degree in Renewable Energies, http://www.emagister.com/master/master-energias-renovables-kwes-1697.htm
Postgraduate Degree in Renewable Energies, http://www.cursosypostgrados.com/programmeas/postgrado-en-energias-renovables-1856.htm
Postgraduate Degree in Renewable Energy Management and Development, http://www.tumaster.com/Postgrado-en-Gestion-y-Desarrollo-de-Energias-Renovables-mmasinfo18529.htm
Postgraduate Degree in Renewable Energies, http://www.mastersadistancia.com/master/postgrado-en-energias-renovables-1856.html
Solar Energy Study Centre Professional Distance Learning Courses, http://www.construmatica.com/formacion/tag/energias_renovables/6
Masters in Renewable Energies, http://postgrado.ceu.es/energias-renovables/
Masters in Environment and Renewable Energies, http://www.escuelademedioambiente.com/pdf/master-medio-ambiente-y-energias-renovables.pdf

Appendix 2: Competencies selected by courses (those competencies which are relevant to the contents that will be offered are marked with a cross)

	Competencies	Sociology of development and global challenges	Environment and sustainable development	Planning of the energy mix	Environmental management and impact evaluation
1.	Capacity for abstraction, analysis and synthesis	X	X	X	
2.	Social responsibility and commitment to citizenship	X	X		X
3.	Ability to use of information and communication technologies			X	X
4.	Commitment to looking after the environment		X		X
5.	Commitment to socio-cultural environment	X			X
6.	Improve and innovate administrative processes using information and communication technologies for the process which allow for its formulation and optimization			X	
7.	Awareness of responsibilities regarding the environment and the values of urban and architectural heritage as well as the capability of knowing and applying research methods to resolve creatively the demands of the human habitat, in different scales and complexities	X	X		X
8.	Ethical commitment regarding the discipline, manifesting social conscience of solidarity and justice, and respect for the environment	X	X	X	X
9.	Ethical commitment regarding the discipline, manifesting social conscience of solidarity and justice, and respect for the environment	X	X	X	X
10.	Ethical commitment regarding the discipline, manifesting social conscience of solidarity and justice, and respect for the environment		X	X	X
11.	Ethical commitment regarding the discipline, manifesting social conscience of solidarity and justice, and respect for the environment	X		X	X

		Courses			
Competencies		Energy and environ-mental policies and regulations	Energy and environ-mental economics	Solar energy	Hydro power
1.	Capacity for abstraction, analysis and synthesis	X			
2.	Social responsibility and commitment to citizenship		X		
3.	Ability to use of information and communication technologies		X	X	X
4.	Commitment to looking after the environment	X			
5.	Commitment to socio-cultural environment	X			
6.	Improve and innovate administrative processes using information and communication technologies for the process which allow for its formulation and optimization	X	X	X	X
7.	Awareness of responsibilities regarding the environment and the values of urban and architectural heritage as well as the capability of knowing and applying research methods to resolve creatively the demands of the human habitat, in different scales and complexities	X			
8.	Ethical commitment regarding the discipline, manifesting social conscience of solidarity and justice, and respect for the environment	X	X		
9.	Provide advice regarding the use of natural resources in the formulation of development policies, norms, plans and programmes, interacting in interdisciplinary and trans-disciplinary areas	X	X	X	X
10.	Development of the professional activity within a framework of responsibility, legality, security and sustainability, planning, executing, managing and supervising projects and services focused in knowledge, exploitation and use of renewable natural resources	X	X	X	X
11.	Propose solutions which contribute to sustainable development, planning, designing and executing research in the topic	X	X	X	X

E-Learning: Sustainability, Environment and Renewable Energy in Latin America 67

	Competencies	Courses				
		Biomass	Wind energy	Energy efficiency and renewable energy	Re energy & project management	Research methodologies focused on sustainability, environ-ment and re
1.	Capacity for abstraction, analysis and synthesis				X	X
2.	Social responsibility and commitment to citizenship					
3.	Ability to use of information and communication technologies	X	X	X		
4.	Commitment to looking after the environment			X		
5.	Commitment to socio-cultural environment					
6.	Improve and innovate administrative processes using information and communication technologies for the process which allow for its formulation and optimization	X	X		X	X
7.	Awareness of responsibilities regarding the environment and the values of urban and architectural heritage as well as the capability of knowing and applying research methods to resolve creatively the demands of the human habitat, in different scales and complexities				X	X
8.	Ethical commitment regarding the discipline, manifesting social conscience of solidarity and justice, and respect for the environment					
9.	Provide advice regarding the use of natural resources in the formulation of development policies, norms, plans and pro-grammes, interacting in interdisciplinary and transdisci-plinary areas	X	X	X	X	X
10.	Development of the professional activity within a framework of responsibility, legality, security and sustainability, planning, executing, managing and supervising projects and services focused in knowledge, exploitation and use of renewable natural resources	X	X	X	X	X
11.	Propose solutions which contribute to sustainable develop-ment, planning, designing and executing research in the topic	X	X	X	X	X

The Challenge of Attracting High-Quality Technology Transfers to Non-BRIC Countries: Chile and its Emerging Wind Energy Industry

A. Pueyo, M. Mendiluce, D. Morales, R. García[1]

Abstract

Literature on the role of technology transfers for the development and deployment of local renewable energy technologies in developing countries often refers to success stories in the so-called BRIC economies. This paper outlines the different challenges faced by a smaller developing country in attracting foreign technologies. Fibrovent, a Chilean company which is entering into the manufacture of wind blades, provides a good case study on how the combination of internal and foreign sources of knowledge, technology push policies and market pull policies can steer the way to high-quality technology transfers, that is, transfers that generate local technological capacity. It also shows some of the barriers that can delay the process, mainly a small and uncertain local demand and the unwillingness of strong potential partners to transfer their technology.

I. Introduction

The United Nations Framework Convention on Climate Change (UNFCCC) considers technology transfers to developing countries as one of the main building blocks of a future international climate change agreement. However, the UNFCCC has so far failed to deliver the rate of TT required to meet the stabilisation challenge. The private sector owns most of the technologies that can facilitate a transition to low-carbon development method but the UNFCCC process appears disconnected from the enabling frameworks which facilitate private investment in emerging economies.

The private sector has responded to the enabling environments and large demand of some emerging economies (mainly Brazil, China and India) where it is already deploying significant amounts of renewable energy technologies. These countries are also assertively pursuing the development of local renewable energy

---

[1] A. Pueyo, *Universidad Politécnica de Madrid*, M. Mendiluce, *WBCSD,* D. Morales and R. García, *CORFO.*

industries. Case study literature has provided detailed accounts of the success stories of BRIC economies[2], including the emergence of leading wind turbine manufacturers in India and China[9, 10, 22], the world-leading Chinese photovoltaic technology[25], and Brazilian biofuels production[26]. However, these countries do not represent the average developing economy. China, India and Brazil offer potential investors the prospect of large profits and have an advanced domestic technological base which allows for rapid adoption of foreign technologies. Smaller developing countries would struggle to replicate BRIC countries' success.

This paper explores the main barriers and drivers to renewable energy technology transfers to Chile. Chile is an interesting case study because of its non-BRIC nature, significant renewable energy resources, stable macroeconomic conditions and new regulations promoting renewable energies. The paper takes stock of the policies implemented by the Chilean government to promote diffusion of renewable energies. The research focuses on the most advanced type of technology transfers, namely those that create indigenous technological capability and local industries. A case study of a Chilean company starting up the manufacture of wind blades is used to reflect the challenges posed by attracting foreign capital and knowledge.

The paper is structured as follows. First, it sets up the international policy framework, highlighting the current gaps in the UNFCCC process for promoting technology transfers to small developing countries and the types of technology push and market pull policies that could be implemented to fill those gaps at the national level. Second, it analyses the barriers and drivers for successful technology transfers in Chile and identifies the policies implemented to support the diffusion of renewable energy technologies. Third, it presents the case study of a specific Chilean company starting-up a business to manufacture blades for wind turbines. The analysis focuses on the internal and external factors that might prevent or enhance a company's ability to absorb foreign technology and to ensure a sufficient demand to grow and recover costs. Finally, the paper concludes with a summary of findings and proposed recommendations for enhancing the level of high quality technology transfers to Chile, highlighting lessons that may also be relevant for other small developing countries.

II. Climate Policy Background

Since its inception, technology transfers (TT) have been an important element of the UNFCCC. The role of TT was enhanced at the 13th Conference of the Parties

---

2 Goldman Sachs coined the term "BRIC countries" referring to Brazil, Russia, India and China as a group of large and fast-growing economies.

(COP) held in Bali in 2007, where technology transfers were identified as one of the four building blocks for a future climate change regime.

Although Copenhagen failed to deliver a binding agreement, it achieved some progress in the field of technology transfer. The Copenhagen Accord proposed a Technology Mechanism (TM) as the basis for subsequent negotiations. Its aim was "to accelerate technology development and transfer in support of action on adaptation and mitigation that will be guided by a country-driven approach and based on national circumstances and priorities"[0]. This new institution was mainly brought forward by developing countries, specifically the G77/China, while Annex I countries preferred using or modifying existing institutions. Financial support was also mentioned in the Copenhagen Accord, with $30 billion in fast-track funding from developed to developing countries for the 2010-2012 period and the creation of a green fund to mobilise $100 billion a year by 2020. However, it was not clear which portion of these funds would be available for TT.

The Cancun Agreements achieved at COP 16 in December 2010 have brought forward the Copenhagen Accord under the auspices of the UN with the approval of the 193 countries working under the Convention. A great deal of work still needs to be done to establish a comprehensive long-term framework for controlling GHG emissions, particularly for the definitions of emission reduction targets and timetables. Even so, significant outcomes have been achieved, including the decision to establish a TM that contains a Technology Executive Committee and a Climate Technology Centre and Network. The objective of the TM is to enhance clean technology development and diffusion. Funding availability for the TM as well as the mechanisms to allocate these funds are still under discussion and eligibility criteria for countries and technologies have not yet been addressed.

To date, the UNFCCC has used three main types of instruments to promote technology transfers to developing countries: the Expert Group on Technology Transfers (EGTT), to be replaced by the Technology Mechanism announced by the Copenhagen Accord; Technology Needs Assessments (TNA), which identify technological gaps of developing countries; and two financial mechanisms: the Clean Development Mechanism (CDM) and the Global Environment Facility (GEF). Neither the CDM nor the GEF were created with the aim of promoting technology transfers, but by channelling private investments or foreign aid towards projects involving clean technologies, they have sometimes indirectly accomplished this goal.

The success of UNFCCC mechanisms in promoting TT has been limited. The EGTT has been criticized for delaying difficult but necessary decisions to enhance TT and for the lack of expertise of its political representatives[2]. TNAs have identified potential projects yet failed to implement them. The GEF only has a limited budget, which has resulted in a lower scale than the market-based CDM.

The CDM has been successful in channelling clean technology investments to developing countries. As of August 2010, 2,447 projects had been registered or were awaiting registration and 2,918 were in the validation process[3]. Annual emissions reductions of projects registered, or awaiting registration, totalled 410 Mt CO_2. Around 36% of CDM projects and 56% of their reduced emissions involved technology transfers, which entails the use of foreign equipment or knowledge not previously widely available in the country[4]. This estimation of TT through the CDM is subject to high uncertainty as it relies on claims which generally are not subjected to a verification test and are primarily based on the developer's own perspective. However, they can provide an idea of the extent of foreign versus local technologies used in the CDM.

Despite of its success in increasing diffusion of clean technologies, the CDM has not been built to the scale required to meet the emission stabilization challenge. Estimated annual financing needs for technology transfers to developing countries are between 150 and $405 billion[6], but the CDM has generated $8-12 billion in annual capital investments[4].

The role of the CDM as a vehicle of TT to developing countries is also questionable, due to its project-based nature, which does not foster large-scale deployment of mitigation technologies or the promotion of innovation in host countries[7]. Most of the projects considered under the CDM scheme are focused on delivering the asset of CERs (certified emission reductions) and were not meant to ensure knowledge development or capacity building for the service supply change which could mitigate greenhouse gas emissions. The same situation commonly occurs with the GEF fund.

Besides, UNFCCC financing has not been able to fill the gaps left by the private sector. Carbon credits via the CDM have facilitated further technology diffusion by reducing the cash back period and increasing project IRR. However, several authors show that successful TT through the CDM is based on pre-existing, enabling frameworks in recipient countries[8, 9]. The CDM has followed the same pattern of geographic distribution used by general FDI flows. China, India and Brazil account for 80% of the emission reductions and 70% of the CDM projects[3]. They are also the main recipients of TT through the CDM, accumulating 80% of the emission reductions and 60% of the CDM projects involving TT (internal calculations based on data collected by S. Seres, E. Haites and K. Murphy, 2009). The attraction of private capital towards low-risk countries with high demand and strong enabling environments is inherent to the markets.

BRIC economies have succeeded in assimilating foreign technologies to build indigenous renewable energy technologies. China and India host leading wind turbine manufacturers which already compete with international, more experienced, suppliers[10], while Brazil is a world leader in the production of biofuels[11]. In addition to their large demand, all of these countries have relied on a set of focused, supportive policies for the diffusion of renewable energy technologies.

Chile has actively participated in the CDM, but the relatively small size of its energy demand compared to larger economies has meant that it only holds 2% of the world CDM projects and 8% of projects in Latin America. In Chile, the implementation of non-conventional/conventional renewable energy (which encompasses all renewable energies except large hydro) is in its first stages of deployment. This is mainly due to the lack of competitiveness of these new technologies compared to conventional sources, and the absence of a stronger policy framework which could encourage renewable energies on their own merit.

A combination of market pull and technology push policies could create stronger enabling conditions for foreign technology transfers and domestic assimilation in Chile.

Technology push policies are policies which influence the supply of new knowledge, increase the absorptive capacity of the recipient country and, therefore, reduce the cost of technology transfer. The most common examples of technology push measures are government sponsored R&D, tax credits for companies to invest in R&D, support for education and training, infrastructure development and funding demonstration projects[12].

Market pull policies influence the demand of technologies which expect cost reductions through a variety of learning processes as the installed capacity increases[13]. Common examples of market pull policies include: tax credits and rebates for consumers of new technologies, intellectual property protection, government procurement, technology mandates, regulatory standards, cap-and-trade schemes, feed-in tariffs, renewable energy portfolios and taxes on competing technologies[12].

III. Preconditions for Renewable Energy Technology Transfer to Chile

A. Economic and Institutional Framework

An enabling environment for investment and the institutionalization of the innovation policy both contribute to the Chilean attractiveness for receiving technology transfers.

The Chilean investment framework is characterized by the country's economic growth, the sharp decline of its public debt, the stabilization of its external accounts and the favourable conditions for international trade agreements, amongst others.

All these factors have contributed to Chile's admission into the selective group of OECD members in 2010, being the first South American member and the first country to join the organization in ten years. In recent years, incoming FDI has maintained an upward trend, helping to increase Chile's competitiveness not only through resources and new markets but also through technological development and specialized knowledge. Over the past decade, FDI has represented an annual average of 6.5% of Chile's GDP, rising to an average of 8% in the 2007-2009 period[14].

There have also been three clear developments in Chile's innovation policy: i) rising consensus for innovation as a development driver and a policy paradigm shift, ii) increasing public support and financial resources for innovation, and iii) progress in institutional strengthening and policy learning[15].

The National Innovation System is responsible for the innovation policy and relies on implementing institutions for enhancing R&D, applied innovation and investment through CONICYT (science and technology council) and CORFO (the Chilean economic development agency). CONICYT covers individual basic research and CORFO covers industrial innovation and applied R&D.

InnovaChile is the implementing agency for CORFO's innovation policies. The agency supports Chilean firms in improving their competitiveness in national and international markets by promoting the development of innovative processes. Its scope ranges from individual companies and networked firms to full production chains, including clusters or geographic groups of companies working in a particular industry. Some of the main beneficiaries include private sector projects and energy-related small and medium-sized companies in cooperation with university research centres.

The High Technology Investment Program (HT) is one of the instruments used by CORFO to attract FDI to Chilean prioritised economic sectors. The instruments used include matching grants for pre-investment, fixed capital, funding start-ups, capacity building and real estate leases.

Recent recommendations suggest that Chilean innovation policy should support the diversification of the economy and the transition towards higher value-added activities[15]. The OECD suggests creating a stronger link between scientific development and industry needs. Therefore, Chile faces the challenge of adapting its current R&D activities and structures to a more strategic approach, which addresses the specific needs of the market. Chile will need foreign knowledge to meet this challenge in many technological sectors.

B. The Chilean Electricity Market

Chile's energy sector is characterized by limited indigenous fossil energy resources, unlike many of its South American neighbours. This situation has reinforced the country's dependence on imported fossil fuels, creating serious periods of electricity shortages during the first decade of this century. On the other hand, Chile's geography is gifted with significant renewable energy potential (mainly hydro resources).

In 1982, Chile was the first country to privatize the electricity market. As a result, investment decisions are based on the marginal cost of electricity production of the available technology portfolio. A reduced, short-term energy price is the main objective over other considerations, such as limiting local greenhouse gas emissions or enhancing local technology development and creating high skilled jobs.

For the past 30 years, Chile's energy policy has been founded on the premise that the best ways to meet the energy demand at affordable prices are to rely on competition between private companies, to regulate natural monopolies and to limit the role of the state in entrepreneurial activities. Embedded in this approach is the assumption that competitive markets will deliver an appropriate level of security of supply. The technology's average cost and reliability are currently the only relevant variables for the grid's expansion, which has led to a heavily concentrated electricity sector and a strong dependence on fossil fuels. Three power companies own more than 90% of the installed capacity at the Central Interconnected System (SIC), while the same situation occurs at the northern grid interconnected system (SING). Long-term planning on capacity expansion to incorporate renewable energies is not easily compatible with the existing model, which is better suited for short-term decisions on dispatch through merit order.

Chilean favourable conditions for FDI and innovation do not mean that foreign low-carbon technologies will be transferred to the country. Instead, the combination of the aforementioned policies could lead to the attraction of energy intensive technologies that will divert from low carbon economic growth.

C. The Status of Renewable Energy Technologies in CHILE

Current installed capacity of renewable energy in Chile is 561 MW, with 171.6 MW of wind capacity. This is equivalent to only 3.7% of the country's total installed capacity. Conditions for the development of non-conventional renewable energy in Chile have improved significantly over the past five years. This has been done through new laws, instruments for direct support to investments (loans and grants for pre-investment studies), better information, the implementation of investment projects and the incorporation of diversification of the energy mix as one of the core objectives of the current energy policy[16].

The regulatory framework for renewable energies has maintained the original goal of minimizing global cost while trying to promote a higher penetration of renewable energies. The main changes include a new market pull policy to encourage non-conventional renewable energy (NCRE) deployment in Chile. Law 20.257, which came into force on April 1, 2008, requires that power companies selling directly to final customers incorporate 5% of NCRE into their electricity sales. This percentage will gradually increase to 10% by 2024. This regulatory amendment applies only to new contracts (therefore, a large percentage of the current contracts are not affected). It is difficult to assess the deadline of all current contracts, and therefore difficult to estimate the future demand of new NCRE projects. This situation creates uncertainty on the real value of the renewable energy attribute.

The renewable energy quota set by Law 20.257 would stimulate a wind power capacity expansion to reach 202 MW by 2020. This estimation considers that the most competitive non-conventional renewable energy (NCRE), particularly hydro, would contribute to most of the target. This means that the law will not trigger a significant increase in the wind power sector. The Chilean Congress is leading efforts to increase the share of NCRE to 20%. If these legal modifications to the current regulatory framework are successful, Chile could achieve nearly 1300 MW of total wind installed capacity by 2020. In any case, these figures contrast the large expected wind capacities of Brazil, with 31.6 GW by 2025, and Mexico, at 6.6 GW by 2025 (IHS, 2010).

Another relevant government effort to enhance renewable energy policies was the creation of the Ministry of Energy in 2009. The Chilean government also created the Centre for Renewable Energy (CER) the same year, with the purpose of supporting the development of a renewable energy industry in the country. The CER activities are focused in two areas: accelerating investment in NCRE and becoming a knowledge and technology transfer hub. The CER is expected to have a lead role in promoting renewable energy technologies in the market and to serve as a clearinghouse for connecting research entities and private companies to the international network of renewable energy technologies.

Since 2005, InnovaChile has supported 102 innovation projects related to the energy sector. It has provided $40 million in funding to develop new technologies such as the production of biofuels from lignocellulosic material and algae (almost 80% of total funding). Some of the beneficiaries include university research centres involved in private sector projects and small and medium-sized companies.

Furthermore, the High Technology Investment Program (HT) opened a new line of promotion for the industry of ancillary services for renewable energy. The programme has successfully attracted five international companies in a single year.

Finally, in 2009 CONICYT and the Ministry of Energy launched a training program on scientific and technological capabilities in the energy sector based on short-term training courses and scholarship for internships. Even though this is a valuable effort, there is no sufficient systematic information on training and human capacity spending for energy R&D in Chile. By the end of 2010, the CER will present a study on the skills and competences required for the renewable energy industry in the country.

IV. The Ministry of Energy does not have a coherent strategy or prioritization of activities to guarantee continued research, development and deployment of new and improved energy technologies[17]. The merit and excellence of individual projects is the primary basis for the approval of basic research funding. As a result, activities are dispersed, collaboration between institutions is lacking, and research is project-driven rather than linked to the country's needs. Renewable energy is not considered a priority sector under the Chilean innovation strategy; rather, it is considered a transversal technology platform relevant to all niches of the Chilean economy. The competitiveness approach might be a narrow goal for future energy R&D priorities, because other factors such as environmental sustainability and energy security should also be taken into account[17]. Case study: development of an indigenous wind component industry in chile.

A. Case Study Description: Chilean Wind Blades Industry

This paper focus on the case study of a partnership between the Chilean company Fibrovent and the Spanish company Eozen, which is expected to formalise soon as a joint venture (JV). The resulting company will be named JV in this paper.

Fibrovent was born in 2001 to produce equipment for the collection of sulphuric acid mist in electro-winning plants in the copper mining industry. The company is part of the SAME Group, which covers a wide range of services for the mining industry. The information gathered in this case study is based on several meetings with the company's executives since November 2008 and a detailed structured interview in August 2010 with the commercial manager of EOZEN Latin America and former manager of Fibrovent.

In 2009, Fibrovent adopted a strategic goal of becoming the first Chilean blade manufacture for the wind power industry, taking advantage of their existing knowledge on composite materials acquired through their operation in the mining industry. This goal is placed within the context of growing international and local demand for wind power technology.

During the process of creating capacity to operate in this new sector, Fibrovent started a partnership with the Spanish wind turbine supplier Eozen. Eozen is a second-tier Spanish wind turbine manufacturer that owns the Vensys generator license. Eozen uses permanent magnet technology, which features a few restricted and high-quality components, and reduced wear parts. It also has a certified blade design for use in Vensys generators.

The objective of the JV is to manufacture blades for 1.5 MW and 2.5 MW wind turbines. Local manufacturing will involve lower logistic and maintenance costs for Chilean wind energy projects. Therefore, Fibrovent is optimistic that this new industrial activity will help make wind power technologies more competitive in the Chilean electricity market.

Initially, the production capacity for the development of the project will be between 75 to 100 MW during 2011. The second phase could expand the production capacity to 150 MW per year in 2012. A third expansion in 2013 would increase capacity to 600 blades per year, or around 300-400 MW.

B. Harnessing Business Opportunities: Potential Demand Size

The relatively small size of the Chilean market presses the JV to grow in international markets. While Latin America is the initial target region, Fibrovent expects to acquire experience to supply to African countries. Competition is considered too strong in the European market, due to the vast number of suppliers already operating in the area. While the United States is a feasible market, due to the Chilean free trade agreement, barriers of entry are higher than South-South commercial operation. Fibrovent expects that the JV wind blades will be able to compete with Chinese products, and plans to penetrate the developing country markets with a high-quality and low-cost product.

However, Fibrovent relies on the local market to be able to grow and compete internationally. The reasons are mainly twofold. Firstly, international recognition will depend on experience. Fibrovent blades need to be operating in a number of wind power projects before they can gain credibility and bid for international projects. Demonstration projects will be much easier to get in the Chilean market. Secondly, Fibrovent lacks the know-how to produce blades which meet high quality standards. It needs foreign partners that can provide the technological knowledge base. While potential foreign partners are not interested in creating an additional competitor in the global market, they are attracted by the possibility of penetrating a new market. The existence of a minimum local market size, as well as the possibility of entering the Latin American market, has been a prerequisite for foreign partners to enter into negotiations for joint production with Fibrovent. International partners expect that the new products provided by Fibrovent will cover the local demand for blades, before other international suppliers can access this market. Therefore, although it has a small potential demand, Chile has been considered an interesting and reliable market for the JV to begin accessing the global market.

C. Taking Advantage of National and International Technology Push and Market Pull Policies

The market conditions required by Fibrovent to start operating in the Chilean wind power industry depend on the implementation of local market pull policies. In the last five years, the Chilean regulatory framework has taken some steps to support renewable energy, but the existing law will only stimulate an additional 30 MW until 2020. This is well below the JV´s ambitions to produce blades for an annual capacity of 400 MW. Although the current administration would like to produce 20% of the country's electricity from NCRE by 2020, there is not yet a clear sign of regulatory modifications to the electricity market which could guarantee a certain level of local demand. Fibrovent considers that policies providing price stability, rather than increasing the price of wind power, would be the most cost-effective way to promote this energy source. Fibrovent is actively involved in local renewable energy policy-making which requests regulatory amendments that extend the demand of renewable energy, including wind power. In any case, given the small size of the Chilean market, the JV relies heavily on global demand.

Technology push policies are also required to increase the knowledge base necessary to implement renewable energy projects once a market has been created for them. An example of these push policies to enhance technology are the incentives for innovation capacity managed by the economic development agency CORFO. Firstly, Fibrovent benefited from CORFO's support by hiring an external consultant from Brazil to advise them on the wind industry. Secondly, to finance the initial capital investments for the construction of the wind blades production plant and the purchase of the necessary equipment. This financing was provided by the programme InvestChile, which was designed to attract foreign investment to Chile. Access to these funds requires a majority participation of foreign investors. This requirement was the main reason Fibrovent decided to structure their new wind blades production business as a JV with a foreign company.

Fibrovent believes that the public sector has not been up to speed in providing the required specialized technical training. Even if Chile has a high general and technical education level, specialized knowledge of the various activities is not available. Training initiatives have included very small amount of technical content. Therefore, knowledge could only be acquired via contact with foreign partners. Fibrovent does not consider the international climate change negotiations as a driver for their investment, but instead as a political framework to enhance the international demand of wind power technologies. Market mechanisms such as the CDM are not seen as a relevant factor for the definition of market size, nor have they provided certainty on future demand.

D. Filling Capability Gaps: Domestic and Foreign Sources of Knowledge

Fibrovent relies on a combination of internal and external sources of knowledge to manufacture wind blades. Through its experience as a service provider for electro-winning processes in the copper mining industry, Fibrovent possesses critical knowledge on composite materials and their properties. Fibrovent has gained experience in advanced technologies and implementing procedures which guarantee high-quality products. The company has built a strong relationship with local suppliers, ensuring the quality of input materials. The steep learning curve to achieve excellence in the mining industry can be now capitalized with the new wind blade business, which requires a strong level of precision and expertise.

The professionalization of human resources is also essential to ensure smooth transition to the new business opportunity. Specialized human capital has increased from the initial sole engineer to the current team of ten engineers, who closely follow international technological developments and are capable of implementing quality management procedures. The quality of general technical education in Chile has been also quoted by the company's members as a key enabler for starting up a new business.

Domestic sources of knowledge are further supported through the company's R&D plan and infrastructure, including its own laboratory on composite materials. The company has also established strong links with local universities to conduct R&D activities. Universidad de La Frontera has been incorporated into the project to preserve the new knowledge created and participate in the design of blades for wind turbines via public grants for basic research.

Previous international exposure is also considered key for a company aiming at meeting international demand. Fibrovent has been exporting glass fibre products for more than five years and has won significant international bids, which have helped it build credibility and self-confidence. Based on previous experience, the company has learned that they need to operate successfully in Chile before competing in the global market.

E. Negotiating a Fair Deal for Technology Transfer

The technology transfer deal is expected to be structured as a joint venture between Fibrovent and the Spanish company Eozen. To date, the deal had not yet been legally formalised. On the one hand, delays in formalisation have been caused by the slow growth of Chilean demand for wind technology and, on the other hand, by the delicate financial situation of the Spanish company Eozen.

In April 2010, Eozen announced a redundancy plan for its 90 employees. In January 2011, new partners entered Eozen, taking 31.33% of Eozen's shares to prevent the company from collapsing³. Negotiations with Fibrovent have therefore been delayed while Eozen's internal situation is stabilised.

Setting up a JV with a foreign technology supplier was a requirement for obtaining InvestChile funds for the initial capital investment. For this reason, Fibrovent did not consider other channels for structuring the foreign technology transfer. InvestChile funds were necessary to reduce the investment risk. The choice of Eozen as a technology partner was motivated by its state-of-the-art technology and its openness to collaboration. The Spanish company was, in fact, the only potential foreign technology provider interested in starting a wind blade production venture in Chile. Other potential partners were interested in penetrating larger markets such as the US, China, India or Brazil. In other cases, wind turbine manufacturers were simply not interested in sharing their knowledge.

Eozen's interest in a joint venture in Chile was motivated by the decline of Spanish demand, where it was not a strong actor. With a weak financial base and a small market share in the declining Spanish market, Eozen needed to expand to high-growth markets in order to survive. A natural expansion region was the Latin American market, due to the cultural affinity and language. Chile was perceived as a low-risk location, with stable macroeconomic conditions, low corruption, and a liberal economy welcoming foreign investors. Recent Chilean legislation to promote renewable energies signalled prospects for wind power demand growth. Not having significant international experience, collaboration with a local company would lower Eozen's entry risks to a new market.

To avoid technology dependence, the initial deal involved the JV's legal ownership of the certified wind blade design as well as training staff on the Spanish manufacturing facilities. During the first stages of their relationship, the foreign technology provider would have the largest contribution. Once operations started, the company would provide a larger contribution by offering lower costs and access to a new market. In their negotiations with the foreign partner, it was essential that Fibrovent had a pre-existing R&D capacity, international exposure, a trained workforce and strong links to the national innovation system (universities and policy makers) and the local suppliers. A potentially large production scale, when bundling the Latin American markets, also increased attractiveness for the foreign technology supplier.

---

3   "Los directivos de la eólica en Andalucía al rescate de Eozen", ALIMARKET Fabricación de Equipos 14 de Enero de 2011.

V. Discussion

A. Sources of Competitive Advantage for the Development of Indigenous Renewable Energy Technologies in Chile

Given its relatively low energy demand compared to other developing countries, Chile would need to specialize in those renewable energies for which it has a comparative advantage, abundant natural resources and previous knowledge.

The analysed case study shows that the mining sector can provide cross-sectoral technology transfers through its internationally competitive suppliers. Knowledge acquired in the mining sector has been transferred to other economic sectors with a notorious level of success. The renewable energy industry is one of these sectors with a high potential for cross-sectoral technology transfer. For example, geothermal exploration requires expensive drilling equipment and human resources for an integral assessment of reservoirs. The mining industry shares a relevant portion of this knowledge and capacities. Concentrated solar power (CSP) technology uses molten salts to store thermal energy during the night in order to provide a continuous supply of power. Most of the salts used in the world's CSP power plants are provided by SQM, a leading Chilean company in the field of fertilizers, lithium, iodine and explosives for mining companies. Wind power technology can also benefit from mining sector expertise, particularly when it comes to the use of copper or composite materials for the manufacture of components.

The case of Fibrovent shows that previous experience in the mining sector had provided technical expertise, credibility in international markets, a competitive and responsive supply chain, and the self-confidence to start a new business and negotiate with potential foreign partners. Links with the national innovation system, mainly technical universities and public innovation institutions, were also an essential element for strengthening their capabilities.

The case study indicates that, before planning the development of indigenous renewable energy technologies, small developing countries should have: highly skilled workforces, local R&D capabilities, and local companies with international credibility. In addition to general technical knowledge, specialized knowledge in the entire value chain of the technologies has been mentioned as a priority. Without these capabilities, negotiation with foreign technology providers could lead to foreign technology dependence. Gaps in the local technology base can be filled with technology push policies, which, in the case of Chile, included grants for universities to undertake R&D projects with industry and public incentives for private innovation in ancillary services for renewable energy technologies.

Other factors that attract foreign technology providers to Chile are: positive macroeconomic indicators, an open economy, low taxes, functioning institutions and a low location risk. These attributes make Chile competitive by lowering transaction costs, exporting costs and labour costs.

B. Barriers to Successful TT

The technology transfer process required for Fibrovent to start producing wind blades in Chile has met with some hurdles, mainly due to a small local demand and the unavailability of strong foreign partners willing to transfer their knowledge and certified blade designs.

Fibrovent's potential foreign partners requested the existence of a sufficiently large local market as a precondition for partnership. Foreign investors are attracted to the large markets of China, Brazil and India and are willing to share their knowledge assets in exchange for the large future profits these markets can offer. Future profits in small developing countries are not as tempting and may not compensate for the loss of competitive advantage involved in relinquishing some knowledge assets. Law 20.257, requiring renewable energy generation quotas, has not been able to stimulate demand of wind power to a significant level and the results of its implementation remain uncertain.

Fibrovent lacked the knowledge to start a new wind blade production business and did not have credibility in the market. It needed a foreign partner who would be willing to enter into a JV in Chile, as this was the requirement to gain access to public funds. Both companies needed each other and the deal has been described as a fair one by Fibrovent's executives. However, Fibrovent could not gain access to a strong technology provider. Its reliance on a financially weaker, second tier supplier has delayed the formalisation of the partnership and created some uncertainties about the feasibility of the venture.

C. Proposals to Overcome Barriers

Small developing countries need to send signals to foreign investors that their market will provide a sufficient capacity to demonstrate local technologies and build credibility. These types of signals are clear in BRIC countries. In India, the Electricity Act of 2003 proposes renewable energy targets and policy support mechanisms, including feed-in-tariffs. In China, the Renewable Energy Law includes a mandatory renewable market share, wind concession programs, premiums for renewable power, VAT relief for renewable energy technologies, emission limits for new vehicles that are stricter than those in the US, and timetables for closure of inefficient production facilities. The Brazilian government began promoting ethanol distilled from sugar cane in the 1970s with the Brazilian Alcohol Program, which triggered technological developments in the automotive industry[19, 11].

In Fibrovent's case, national demand-pull policies have not been strong enough to ensure the quick-start of its project and to raise interest among a wider group of potential partners. A more effective implementation of the renewable energy quota, as well as policies providing price stability, could boost wind energy demand. Price stabilisation, rather than subsidies through feed-in tariffs, is more likely to work in Chile, given the high price of Chilean electricity and its free market economy. The CER is currently studying different potential mechanisms for price stabilisation.

As regards international demand-pull policies, the Clean Development Mechanism (CDM) has not played a significant role in EA's business plan, which shows the need to create international instruments that can effectively address low carbon innovation in developing countries.

Once credibility is gained through operations in the local market, the market pull element will mainly come from international markets. The JV business plan is designed to export most of the wind blades produced. Access to the Latin American market through a regional base in Chile would be an important first step towards internationalisation of the JV's operations and presents a number of advantages for foreign partners. Firstly, being a small country facilitates the formation of alliances because there is no fear of creating heavy global competition. Strong protection of intellectual property rights (IPR) also provides the right signals for foreign knowledge-intensive industries fearing loss of competitiveness as a result of sharing knowledge with low-cost manufacturing companies in developing countries. In addition, Chile's smaller size gives technology providers the flexibility to adapt to the market's varying conditions and to recover more quickly from an incorrect design or marketing decisions before approaching the global market.

Fast, international competitiveness is more urgent in small developing countries which do not have the bargaining power of big economies to protect their infant industries. Local content requirements, as implemented in China[9, 10, 22], would be highly inefficient in Chile, where there is no local industrial available to provide the required components. The lack of local capacity would make this policy highly inefficient and make the Chilean market unattractive for foreign suppliers. Instead, Chile could build upon its open market philosophy, removing barriers for foreign entry while rewarding local sourcing through, for example, a favourable fiscal treatment. Fibrovent believes that international competitiveness can be achieved through temporary, favourable treatment of local suppliers, as permanent protection may lead to poor performance.

VI. As regards barriers due to lack of strong partners, our case study has shown that the requirement of a JV to access InvestChile funds may have delayed the TT process. The Government justifies majority foreign participation by the need to transfer technologies. However, this condition limits the opportunities for local innovators relying on foreign licenses and in-house learning processes for TT. Technology push policies similar to Invest but aimed at local innovators could benefit local companies which have significant absorptive capacity, by reducing their financial dependence on foreign companies. Conclusions.

Literature on the role of technology transfer for the development and deployment of local renewable energy technologies in developing countries often refers to success stories in the so-called BRIC economies. This paper has shed some light on the different challenges faced by smaller developing countries. The specific case of the Chilean emerging wind power component industry has been used as an example.

Non-BRIC developing countries are highly heterogeneous and the lessons learnt through the case of the emerging Chilean wind power industry may not be replicated in all of them. Chile´s free trade philosophy, skilled technical workforce, ease of doing business and macroeconomic stability make it a unique example. In any case, some of the lessons learnt are relevant for all policy makers in non-BRIC developing countries:

1. The importance of focusing on a set of prioritized technologies to be able to create a critical mass of highly-skilled workers and installed capacity
2. The convenience of focusing on technologies which have a pre-existing knowledge base from economic sectors which are already competitive, such as the mining industry in Chile
3. The importance of a minimum local demand and the access to foreign markets to attract foreign technology providers. Effective national market pull policies depend highly on each country's circumstances. Chile's liberal economy and high electricity prices made it choose renewable energy quotas as the most cost-effective policy. However, some flaws in the implementation of legislation and the uncertainty of future electricity prices have prevented the growth of renewable energy and deterred TT activities.
4. The definition of flexible technology push policies to attract foreign technology, which do not prescribe the channel to structure the transfer. For example, when local capabilities are high, licensing or importing equipment, together with profound internal learning processes, could be more appropriate than channels which create a stronger dependence on foreign technology providers.

Further research would be necessary to describe the experiences of other non-BRIC developing countries'.

VII. References

[1] UNFCCC (2009). 15th session of the Conference of Parties (COP 15) to the United Nations Framework Convention on Climate Change (18 December 2009), http://unfccc.int/resource/docs/2009/cop15/eng/l07.pdf.

[2] South Center and Center for International Environmental Law (2008). The technology transfer debate in the UNFCCC: politics, patents and confusion, Intellectual Property Quarterly.

[3] UNEP Risoe (2010). CDM Pipeline 01-08-10.

[4] S. Seres (2008). Analysis of Technology Transfer in CDM Projects.

[5] UNFCCC (2008). Investment and financial flows to address climate change: an update.

[6] McKinsey & Company (2009). Pathways to a Low-Carbon Economy: Version 2 of the global greenhouse gas abatement cost curve.

[7] B.C. Staley and C. Freeman (2009). Tick Tech Tick Tech: Coming to Agreement on Technology in the Countdown to Copenhagen. Washington, DC, World Resources Institute.

[8] U.E. Hansen (2008). Technology and knowledge transfer from Annex I countries to non-Annex I countries under the Kyoto Protocol's Clean Development Mechanism (CDM) – An empirical case study of CDM projects implemented in Malaysia. Department of Geography and Geology. Copenhagen, University of Copenhagen. Master's thesis.

[9] B. Wang (2009). Can the CDM bring technology transfer to developing countries? An empirical study of technology transfer in China's CDM projects. The Governance of Clean Development. U. o. E. Anglia. Working paper 002.

[10] J. Lewis (2007). "Technology Acquisition and Innovation in the Developing World: Wind Turbine Development in China and India." Studies in Comparative International Development (SCID) 42(3): 208-232.

[11] S.K. Ribeiro and A.A. De Abreu (2008). "Brazilian transport initiatives with GHG reductions co-benefit." Climate Policy 8(2): 220-240.

[12] G.F. Nemet (2007). Policy and Innovation in Low-Carbon Energy Technologies. Energy and Resources, University of California, Berkeley.

[13] M. Grubb (2004). "Technology, innovation and climate change policy: an overview of issues and options." Keio Economic Studies 41(2): 103-132.

[14] Chilean Foreign Investment Committee (2009). http://www.foreigninvestment.cl

[15] OECD (2009). Strengthening institutional capacities for innovation policy design and implementation in Chile. http://biblioteca.cnic.cl/media/users/3/181868/files/18813/STRENGTHENING_INSTITUTIONAL_CAPACITI

ES_FOR_INNOVATION_POLICY_DESIGN_AND_IMPLEMENTATION_IN_CHILE.pdf.
[16] CNE (2009). Non-Conventional Renewable Energy in the Chilean Electricity Market. http://www.investchile.com/incjs/download.aspx?glb_cod_nodo=20080827173707&hdd_nom_archivo=Non-Conventional_Renewable_Energy_in_the_Chilean_Electricity_Market.pdf
[17] International Energy Agency (2009). "Chile Energy Policy Review".
[18] CONAMA, Environmental Impact Assessment System (SEIA). August 2010. www.seia.cl
[19] D.G. Ockwell, J. Watson, G. MacKerron, P. Pal and F. Yamin (2008). "Key policy considerations for facilitating low carbon technology transfer to developing countries", Energy Policy 36(11): 4104-4115.
[20] W. Cai, C. Wang, W. Liu, Z. Mao, H. Yu and J. Chen (2009). "Sectoral analysis for international technology development and transfer: Cases of coal-fired power generation, cement and aluminium in China", Energy Policy 37(6): 2283-2291.
[21] H. Machado (2009). "Brazilian low-carbon transportation policies: opportunities for international support", Climate Policy 9(5): 495-507.
[22] X.L. Zhang, S.Y. Chang, M.L. Huo and R.S. Wang (2009). "China's wind industry: policy lessons for domestic government interventions and international support", Climate Policy 9(5): 553-564.
[23] Victor Poblete (2010). Interview held in Santiago, Chile in July 2010.
[24] S. Seres, E. Haites and K. Murphy (2009). "Analysis of technology transfer in CDM projects: an update", Energy Policy 37(11): 4919-4926
[25] A. de la Tour, M. Glachant and Y. Ménière (2010). Innovation and international technology transfer: the case of the Chinese photovoltaic industry. CERNA Working Paper Series. Paris, Cerna, Centre d'économie industrielle, MINES Paris Tech.
[26] A. Hira and L.G. de Oliveira (2009). No substitute for oil? How Brazil developed its ethanol industry, Energy Policy 37(6): 2450-2456.

Fostering Renewable Energies in Small Developing Island States Through Knowledge and Technology Transfer: Findings from a Labour Market Survey Undertaken in Mauritius under the DIREKT Project

V. Schulte[1], D. Surroop[2], R. Mohee[2], P. Khadoo[2], W. Leal Filho[1], J. Gottwald[1]

Abstract

Renewable energy is of major importance for sustainable socio-economic development in the ACP (Africa, Caribbean, Pacific) small island developing states (SIDS) for two main reasons. Firstly, due to the obvious environmental benefits of reduced or no CO_2 emissions. Secondly, because it offers local job opportunities and contributes towards reducing the dependency on imported fuels. Renewable energy goals are on the political agenda of many small island developing states, some of which have ambitious plans. However, the current potential is not used despite the high availability of natural resources. One of the main problems is the lack of policies combined with logistical barriers, such as the lack of local expertise available to plan, design, implement and maintain renewable energy technologies.

Due to the innovative nature of this field, institutions of higher education are very important actors in this sector, especially in terms of research as well as educating future employees. Despite the value of the topic of renewable energies, it is not as yet prominently featured in the curriculum and research activities of universities in Europe and in the ACP region as should be the case.

In order to address this perceived need, the Small Developing Island Renewable Energy Knowledge and Technology Transfer Network (DIREKT project), funded by the EU ACP Science and Technology Programme, is being undertaken. The purpose of the project is not only to improve the academic quality of higher education institutions in the ACP region, but also to strengthen their role in contributing towards innovation, local economic development and social cohesion.

---

1 Research and Transfer Centre "Applications of Life Sciences", Faculty of Life Sciences, Hamburg University of Applied Sciences.
2 Chemical and Environmental Engineering, Faculty of Engineering, University of Mauritius.

This paper introduces the DIREKT project and presents the key findings of an assessment needs analysis on the potential of renewable energy technologies in Mauritius. The objectives of these surveys were to identify the needs of the labour market regarding education and research in the renewable energy sector, to identify training needs of university staff in the renewable energy sector and to benchmark renewable energy activities among higher education institutions. The findings showed that most organizations would like to have in-house training and information, services and opportunities offered by tertiary institutions, which can be in the form of seminars or workshops. Most businesses also think that implementing photovoltaic systems would be the most appropriate renewable energy (RE) technology in Mauritius.

I. Introduction

For some time now, ample sources of oil, coal, natural gas and nuclear fuel have provided the driving force behind growth in the economy as a whole. At the dawn of the twenty-first century, two major factors have changed this paradigm. The first is the awakening of large-scale demand for energy in the new market economies. The second is the accelerating depletion of global oil reserves coupled with a lack of sufficient new discoveries that can both provide a replacement for the depleted reserves and meet the added demand. This crisis has led to the decision to focus on implementing sustainable energy programmes at various levels worldwide.

Mauritius comprises a main island of 1,870 km^2 800 km off the east coast of Madagascar at a latitude of 20° south and a longitude of 58° east as well as several surrounding islands, all of which are volcanic in origin and encircled by fringing coral reefs that enclose lagoons of various sizes. The climate is subtropical with winter prevailing from May to September and summer from October to April. Mauritius has no known oil, natural gas or coal reserves, and therefore depends on imported petroleum products to meet most of its energy requirements. 80% of its energy is derived from burning imported fossil fuels, which is costly and has increased green-house gas emissions by 20% since 2000.

Mauritius is presently a member of the DIREKT project where an assessment of the potential renewable energies was conducted. The Small Developing Island Renewable Energy Knowledge and Technology Transfer Network (DIREKT) is a collaboration scheme involving universities from Germany, Fiji, Mauritius, Barbados and Trinidad and Tobago. It aims to strengthen the science and technology capacity in the field of renewable energies in a sample of ACP (African, Caribbean, Pacific) small island developing states by focusing on technology transfer, information exchange and networking. Developing countries are especially vulnerable to problems associated with climate change and much can be gained by raising their capacities in the field of renewable energies, which is a key area. The project is funded by the ACP Science and Technology Programme and EU programme for cooperation between the European Union and the ACP region.

Renewable energies are presently of great relevance for the socio-economic development of countries in the ACP region (particularly in small island developing states) as well as in Europe, as to date they heavily depend on (imported) fossil fuels to meet their energy needs. Apart from the environmental benefits and the fact that it concretely contributes to mitigating climate change, the local generation and use of renewable energies offer great potential for local economic development. There are several studies that have been conducted globally to promote renewable energies. Shen[1] assessed the 3E (energy, environment and economy) goals and renewable energy regulations. Karekezi[2] evaluated the potential of renewable energy in sub-Saharan Africa. Balat[3] studied the contribution of green energy sources to electrical power production in Turkey. Johansson[4] examined the issue of renewable energy and policy options in Europe.

A study was therefore initiated under the DIREKT project to conduct an assessment need analysis on the potential of renewable energy technologies in Mauritius. The assessment need analysis was based on a survey conducted in the business sector and among other organizations in order to determine the most likely source of renewable energy that is suitable on the island.

II. Overview

A. Energy Sector Overview of Mauritius

Mauritius constitutes many organizations which are responsible for catering for energy on the island. The Ministry of Renewable Energy and Public Utilities (MPU) is responsible for energy policy and its portfolio includes energy, water and wastewater. The Central Electricity Board (CEB), which was established in 1952, is empowered to prepare and carry out development schemes with the general object of promoting, coordinating and improving the generation, transmission, distribution and sale of electricity in Mauritius. The CEB produces around 40% of the country's total power requirements from its four thermal power stations and eight hydroelectric plants, the remaining 60% being purchased from independent power producers, mainly sugar industry-owned, who produce electricity from bagasse/coal.

B. Electricity Production in Mauritius

A major part of the fuel imported goes to the production of electricity, whilst the rest is shared among the transport, manufacturing, commercial and private sectors. According to the Central Electricity Board (CEB) Annual Report[5], the total electricity generated in 2008 was 2,199 GWh and the electricity consumption 2,054 GWh. Of the total electricity produced, 44.2% was produced by CEB generating facilities and the remaining 55.8% by the Independent Power Producers (IPP) of the sugar industry, burning bagasse and/or coal as shown in Table 1.

Table 1: Electricity Generation in Mauritius in 2008

	GWh	%
CEB		
Thermal Diesel/Gas Turbine	888.40	40.40
Hydro	83.86	3.80
IPP		
Bagasse/Coal	1182.65	53.80
Other/Hydro	44.01	2.00
Total	2198.92	100.0

(Source: CEB, 2008)

The combined bagasse/coal steam power station produced a total of 1,182.65 GWh in 2008, of which 447.52 GWh from bagasse and the remaining 735.13 GWh from coal[5].

C. Targets for Renewable Energy Between 2010 and 2025

Based on government overall energy policy and strategy and renewable and non-renewable sources of energy, the targets in terms of percentage of total electricity generated between 2010 and 2025[6] are provided in Table 2 below.

Table 2: Forecasted energy mix

Fuel Source		Percentage of Total Electricity Generation			
		2010	2015	2020	2025
Renewable	Bagasse	16%	13%	14%	17%
	Hydro	4%	3%	3%	2%
	Waste to energy	0	5%	4%	4%
	Solar PV	0	2%	6%	8%
	Wind	0	1%	1%	2%
	Geothermal	0	0	0	2%
	Sub-total	20%	24%	28%	35%
Non-Renewable	Fuel Oil	37%	31%	28%	25%
	Coal	43%	45%	44%	40%
	Sub-total	80%	76%	72%	64%
	Total	100%	100%	100%	100%

Note: Forecasted energy mix 2010-2025

III. Methodology

This section presents the methodology adopted in this study to carry out the survey. The purpose is to show in detail the process for designing the questionnaire that was used for the research work. It discusses the details of the research, including the problem formulation, research design, sample design and size. This study was designed to determine the extent to which private and public sectors have implemented or are ready to invest in renewable energy technologies in Mauritius. To measure the effectiveness, both quantitative and qualitative data have been used.

This study is primarily designed to determine the degree to which renewable energy technologies have been implemented to date in Mauritius and whether or not stakeholders plan to implement these technologies in the future. For the purpose of this project, the primary concern of the population size based on those involved in the renewable energy sectors. As a result, a random selection process was used to build a survey population (i.e. the people who will participate in the survey). All organizations and institutions throughout Mauritius which are in one way or another involved in renewable energies were identified. Through preliminary screening, the most appropriate companies and institutions were selected. In this respect, 45 organizations and institutions were targeted for this study, including universities, research institutions, ministries, utility companies and private companies involved in renewable energy technologies and NGOs, among others.

Data was collected using a questionnaire over a period of two months. Questionnaires were sent to the targeted groups by post, email and fax, whilst others were delivered in person. The informal face-to-face interview was also used with some of the target groups due to the nature and content of the questionnaire. Appointments were arranged with the directors, managers and employees by phone and email one week before. Each face-to-face interview lasted approximately 15 minutes. The face-to-face method provided an opportunity to dig further (i.e. ask for addition information) if the answers given by the respondents were incomplete or ambiguous. Furthermore, it provided rich and descriptive data. Some filled-in questionnaires were returned by post while others returned by email. Trends in solar and wind speed were observed at meteorological stations for a period of three months to gain an in-depth understanding of the weather trends in Mauritius.

The returned questionnaires were scanned individually immediately after they were received. All completed questionnaires were considered usable, as they had been filled in correctly. The SPSS analyser program was used to analyse the data obtained. Since the survey aimed at obtaining a lot of quantitative data, tables and figures were used to present the information obtained in an understandable manner and their significance appreciated.

IV. Analysis and findings

A. Presentation of Findings

The survey is divided into the following parts:

Part 1: About the organizations
Part 2: Research and innovation needs of the organizations
Part 3: Staff training needs of the organizations

B. Part 1: About the Organizations

1. Name and Type of Organization

The idea behind this question was to find out basic information about the organization such as their name, type and year of establishment.

2. Number of Employees in the Organization

The objective of this question was to determine the number of staff in the organization and those involved in the renewable energy sectors. The outcomes of the survey are shown in Table 3. The table below reveals that, at a majority of the companies, more than 30% of the employees are active in renewable energy fields. Moreover, at the NGOs Environment Protection & Conservation Organization and Le Cercle D'Epanouissement Féminin, 100% of the staff are involved in renewable energies, which implies that these organizations have been set up to work exclusively towards renewable energies.

Table 3: Number of employees in the organization

Business and Organizations	Total no. of employees in Organization	Number of employees in Renewable Energy Area	Percentage (%) of employees in Renewable Area
Waste Water Management Authority	150	5	3.33
Environment Protection & Conservation Organization	4	4	100.00
Le Cercle D'Epanouissement Feminin (NGO)	2	2	100.00
Mauritas	7	3	42.86
Ecofuel Limited	70	7	10.00
Sotratech Limited	64	12	18.75
Ministry of Renewable Energy	50	30	60.00
Aquaflo Limited	14	5	35.71
Energy Services Division	241	21	8.71
Falcon Citizen League	11	4	36.36
Maurice Ile Durable Fund	4	2	50.00
Ministry of Environment and Sustainable Development	35	20	57.14
Central Water Authority	1541	13	0.84
Central Electricity Board	1897	80	4.22
Ministry of Energy and Public Utilities	50	3	6.00
Green zone Limited	10	6	60.00
Megasun	14	8	57.14
Solartech Company Limited	9	4	44.44
Extreme solar	12	8	66.67
British American Investment (Energy division)	8	6	75.00
OMNICANE	489	137	28.02
Environment Care Association	7	4	57.14
MAUDESCO	16	8	50.00
Land Based Oceanic Industry	34	18	52.94
Green Sun	11	5	45.45
Total	4866	388	

3. Renewable Energy Technologies that Interest Organizations

This question addresses the types of renewable energy technologies that interest organizations or they have already implemented. It can be deduced from the chart below that most organization are very much interested in photovoltaics

(56%) followed by solar thermal energy (52%), wind power (48%), hybrid systems (40%), biomass (24%), biogas (24%), biofuel (16%), hydropower(16%), geothermal (12%) and ocean energy (12%). The objective of this question was to determine interest in renewable energies and the most appropriate renewable energy technologies for future application in Mauritius. The survey revealed that stakeholders and organizations are mostly interested in solar and wind energy technologies.

Fig. 1: Renewable energy technologies that interest organizations

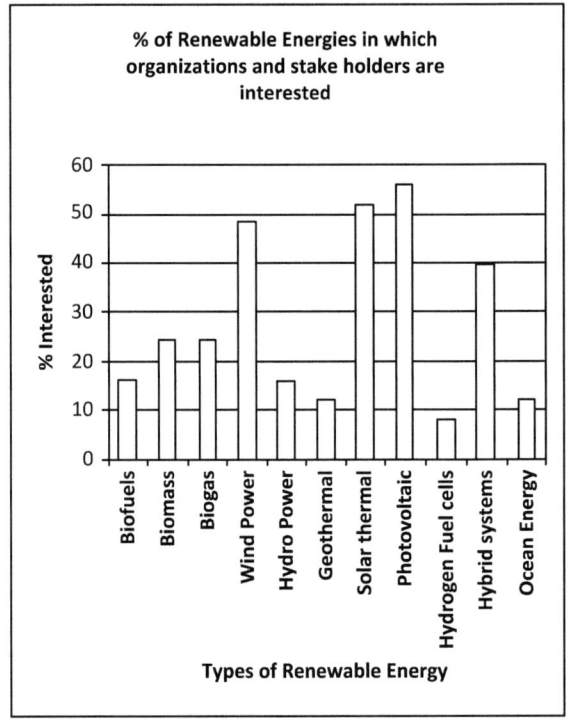

C. Part 2: Research and Innovation Needs of the Organizations

1. Ability to Meet Future Research/Innovation Needs

This part consisted of analysing the ability of the various organizations to meet future research and innovation needs set by the government. The charts reveal that 40% of the targeted group has a satisfactory and good ability to design and produce new renewable energy products and systems for specific users when and where required and 56% of the organizations can carry out renewable energy resource assessment to a satisfactory level. Regarding the gathering of renewable energy (RE) data from established sources, only 52% of the targeted groups have a good level. However, 40% of the group is only fairly satisfactorily able to evaluate the economics on RE technologies and 24% are able to manage RE projects. As regards writing of funding proposals for RE projects, only 40% of the organizations have a good level of ability. As a whole, it can be concluded that the majority of the group is able to deal with future innovations and research needs though some efforts have to be made in the fields of evaluating economics and managing RE projects.

Fig. 2: Design and produce renewable energy products

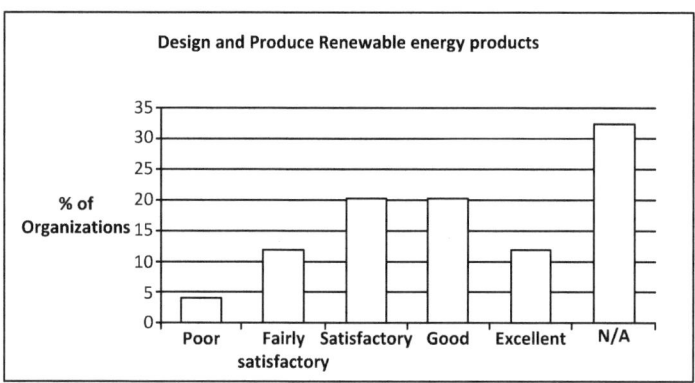

Fig. 3: Renewable energy data

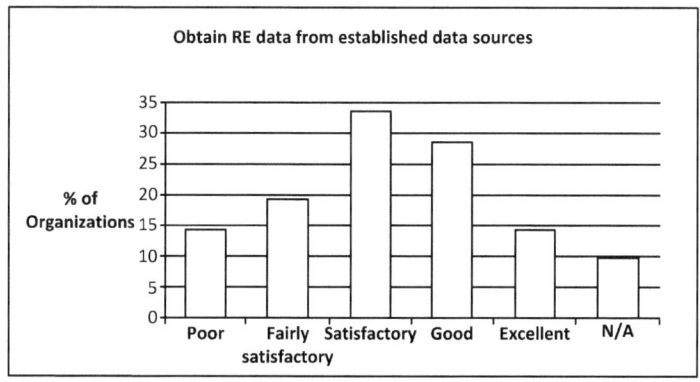

Fig. 4: Economics of RET

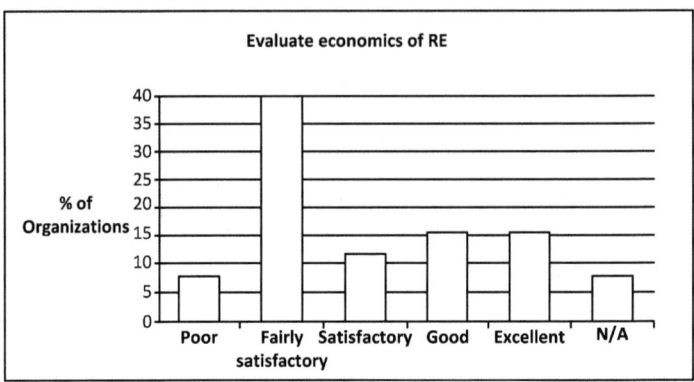

Fig. 5: RE projects

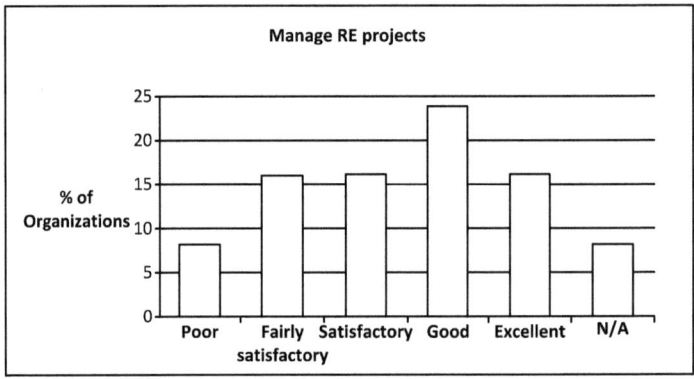

Fig. 6: Writing funding proposals for RE projects

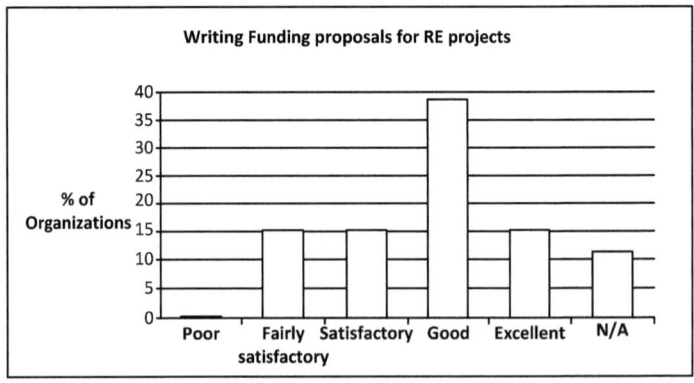

Fig. 7: Carry out renewable energy resource assessment

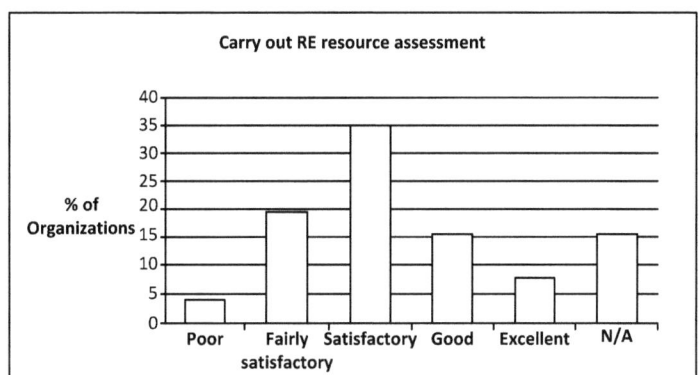

2. Incentive Schemes for Renewable Energy Business

This question aims to find the level of knowledge the respondents have of government and other assistance and incentive schemes for RE businesses.

From the bar chart, it can be inferred that 84% of the targeted group are well aware of the incentive schemes set by the government on RE. As expected, the result shows that those organizations which are involved in RE are well aware of all the benefits that using and promoting RE will bring to the economy and the environment of Mauritius.

Fig. 8: Level of knowledge about RE business

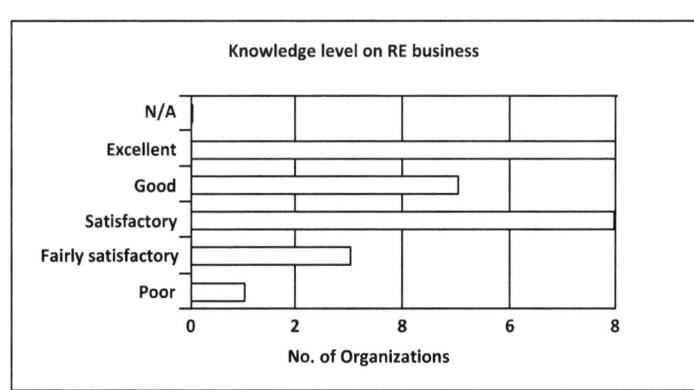

3. Benefits of the Services Provided by Tertiary Institutions

This question was asked in order to determine which of the information or services, if provided by universities or technical institutes, would be most beneficial or least beneficial to the organizations. The outcome of the survey showed that the targeted groups found that information on evaluating economics of RET, design, construction and installation, and seminars, workshops and short courses would be of utmost importance to them and training in project management would also benefit their organization. None of the organizations found that the proposed services would not be beneficial to them.

Fig. 9: Benefits of information and services

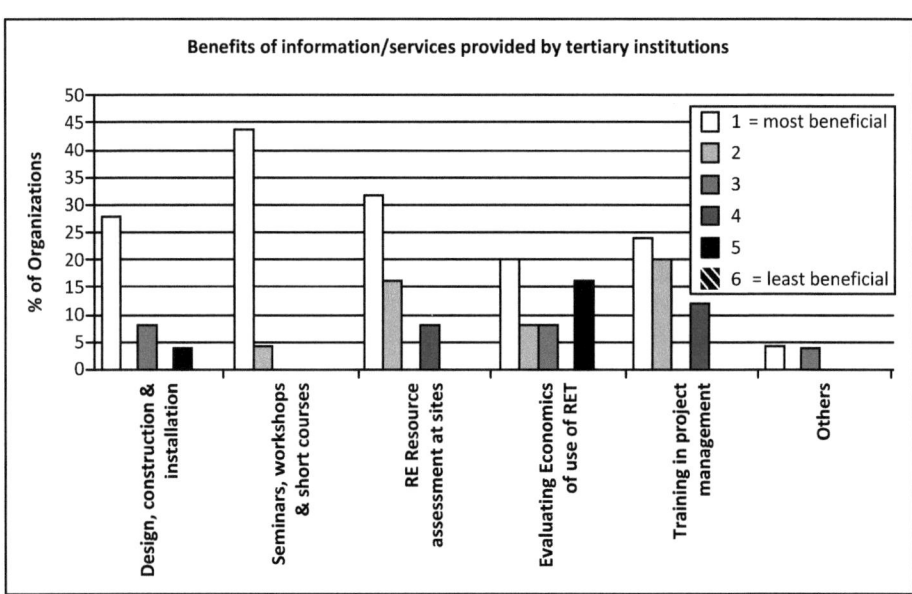

4. Services Offered by Tertiary Institutions

This question identifies which types of services/opportunities that can be offered by tertiary institutions would the targeted group be most interested in. The results are shown in Figure 10.

Fig. 10: Types of services

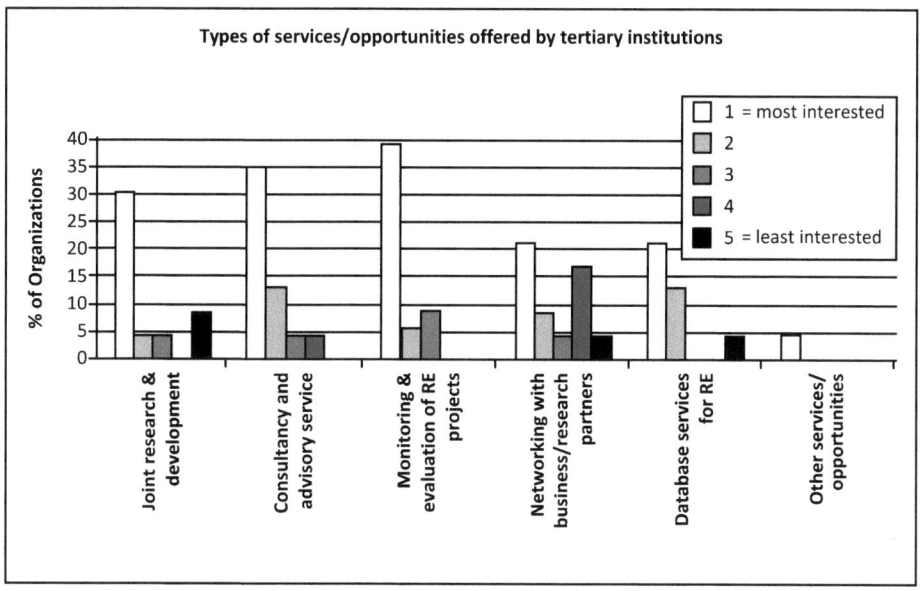

91% of the targeted groups were mostly interested in consultancy and advisory service and monitoring and evaluation of RE projects. Only 4% of the group specified other services which were not mentioned.

5. Market-Oriented Service Expected from Tertiary Institutions

This question assesses the type of market-oriented services that the organizations would prefer to receive from tertiary institutions. The responses obtained were very useful and included training seesions for putting RE strategies into practice, international organizations providing information related to funding grants for RE projects and partnership with private consulting firms. Moreover, many organizations are very much interested in receiving training in construction and installation of RE devices mainly in the field of wind and solar energy and to undertake international and regional market surveys. Concern among the targeted group was also diverted towards carrying out feasibility studies and having complete assessment of sites in terms of RE sources up to implementation of project, including installation, testing and commissioning. The sample results show that the organizations are very much willing to be well informed and guided on the RE technologies to the extent of installation and commissioning.

D. Part 3: Staff Training Needs of the Organizations

1. Knowledge of Renewable Energy Technology

The purpose of asking this question was to analyse the extent to which staff of the targeted group are aware of RE resources and technologies.

Interestingly, the majority of managers (72%) have a good level of general awareness of RE, 44% have adequate academic training in RE science and technology amongst their staff, 48% are academically trained in RE management, and 52% of the targeted group's staff have had previous work experience in the field of RE. It must be noted that a high percentage (68%) of the finance staff do not have academic training in RE finance. Based on this, tertiary institutions can devise a training service or course which would be beneficial to the finance staff in terms of RE practices. Moreover, though not substantial, around 8% of the targeted group's staff have acquired knowledge and know-how via other means such as seminars and technical training sessions as well as suppliers and foreign universities.

Fig. 11: Level of RE knowledge

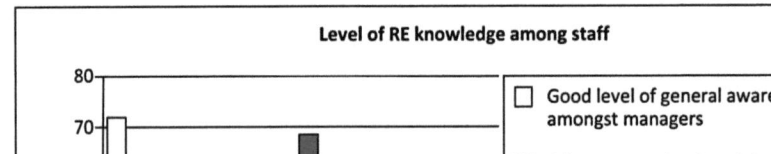

2. Training Requirements

This question aims to assess the importance of the different training courses for the staff in different organizations. The results obtained are displayed in Figure 12.

56% of the organizations suggested that an in-house training session would be very beneficial for their staff, while 32% considered formal technical education and learning through job experience to be most important. 28% agreed that seminars are important for the development of their staff in the RE sector. However, 16% of the organizations found that formal university education was not necessary for the smooth running and understanding of appropriate RE technologies by their staff.

Fig. 12: Most appropriate training for staff

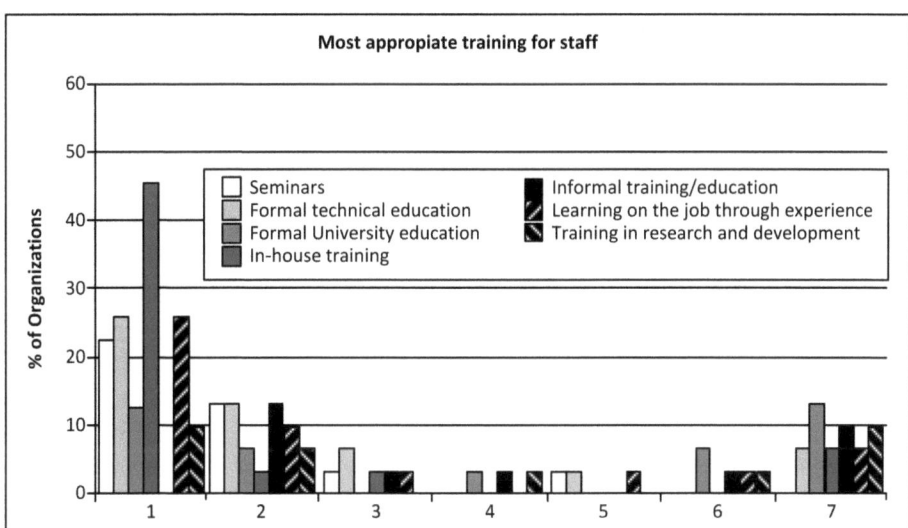

(1 = most important, 7 = least important)

V. Conclusion

From the results obtained, it can be concluded that organizations and businesses are mostly interested in photovoltaic technologies, followed by wind power and a hybrid of wind and solar technologies. It can also be concluded from the results that organizations are more interested in in-house training and learning on the job through experience rather than formal technical and university education. The survey also showed that organizations are capable of carrying out renewable energy resource assessments and obtaining data from other data sources. But to a large extent, organizations lack the ability to manage and evaluate economics of RE projects. Consequently, the majority of the organizations are interested in receiving information, services and opportunities in the form of seminars or workshops provided by tertiary institutions.

VI. Acknowledgement

The authors would like to thank the EU ACP Science and Technology Programme for funding this study under the ACP support scheme.

VII. References

[1] Y.C. Shen, T.R.L. Grace, L. Kuang-Pin and B.J. Yuan (2010). An assessment of exploiting renewable energy sources with concerns of policy and technology. Energy Policy, 38: 4604-4616.
[2] S. Karekezi (2002). Renewables in Africa – meeting the energy needs of the poor. Energy Policy 30: 1059-1069.
[3] H. Balat (2008). Contribution of green energy sources to electrical power production of Turkey: A review. Renewable and Sustainable Energy Reviews, 12: 1652-1666.
[4] T.B. Johansson and W. Turkenburg (2004). Policies for renewable energy in the European Union and its member states: an overview. Energy for Sustainable Development, vol. VIII, no. 1.
[5] Central Electricity Board (CEB). Annual Report. 2008.
[6] Ministry of Public Utility (2009). Long term energy strategy 2010-2025. Report. 2009.

Defining a Mitigation Strategy in a Developing Country Context: The Case of Chile

R. O'Ryan, M. Díaz, J. Clerc[1]

Abstract

In this paper, we provide a forecast of the evolution of greenhouse gas (GHG) emissions in Chile and evaluate alternative policies and specific instruments to deal with future contingencies the country may face in this area. We focus exclusively on emissions resulting from fossil fuel combustion in stationary and mobile sources, and do not consider emissions from industrial processes or from changes in land use and forestry. We identify and assess the main measures for reducing greenhouse gases in the transportation, commercial, public and residential, industrial, mining, and electricity-generating sectors. We also define economic and regulatory instruments which could help to promote and implement proposed measures.

I. Introduction

In this paper, we provide a forecast of the evolution of GHG emissions in Chile and evaluate alternative policies and specific instruments for dealing with future contingencies the country might face in this area. We focus exclusively on emissions resulting from fossil fuel combustion in stationary and mobile sources, and do not consider emissions from industrial processes or from changes in land use and forestry.

We identify and assess the main measures for reducing greenhouse gases in the transportation, commercial, public and residential, industrial, mining, and electricity-generating sectors. We also define economic and regulatory instruments which could help to promote and implement proposed measures.

For each abatement option, implementation costs and reduction potentials are estimated utilising expert opinion, company and regulatory agency information, and information gathered through review of national and international literature. In all cases, the estimated cost of the measures reflects the real costs to the society of adopting them.

---

1 University of Chile.

Taking all of the above factors into account, we created a GHG emissions abatement curve for Chile, which shows the reduction potentials and costs of the different measures assessed. The curve is ordered from the least expensive (measures with negative costs) to the most expensive per ton of CO_2e reduction, and the total cost of the maximum possible potential.

We constructed different scenarios to analyse emission reductions. The first scenario is a baseline which considers current energy consumption and emission trends. Two GHG emissions reduction scenarios for Chile are developed accordingly to combine the potential reductions available through the different measures. The scenarios correspond to the so-called "early actions" (actions that have been carried out in Chile in recent months) and the maximum possible emissions reduction, which includes the aggregate interactions of the majority of the measures. This is not a straightforward and simple exercise, as in no case does the result equal the sum of the reductions of each separate measure. The Long Range Energy Alternative Planning (LEAP) system model was used to allow the different variables to interact in equilibrium in accordance with the projected energy demand for each sector.

In conclusion, the foundations for a Chilean policy on GHG emissions reduction are presented. The measures which would be most appropriate to implement, based on their cost effectiveness, are reviewed along with tools to make implementation feasible in a developing context. Finally, how to consider all of the above in a proposed climate change policy which is realistic for the country in the international context of climate change and takes into account the economic implications which could stem from the country's stance towards the United Nations Framework Convention on Climate Change (CMNUCC) is discussed.

II. Evolution of Greenhouse Gas Emissions and Relevant Indicators

This section describes the evolution of GHG emissions and provides a very brief analysis of how certain indicators related to energy consumption and CO_2 emissions evolved from 1990 to 2006.

The Long Range Energy Alternative Planning (LEAP) system model was used to estimate emissions from energy consumption. This tool is one of the most widely used models in the world, with hundreds of users in more than 140 countries, and has served as a means of communicating with the IPCC. The model basically uses the IPCC's emission factors and allows for the structuring of information according to the different end-use consumer sectors. The results are presented in the following figure.

Fig. 1: Evolution of GHG emissions by sector, 1986-2006

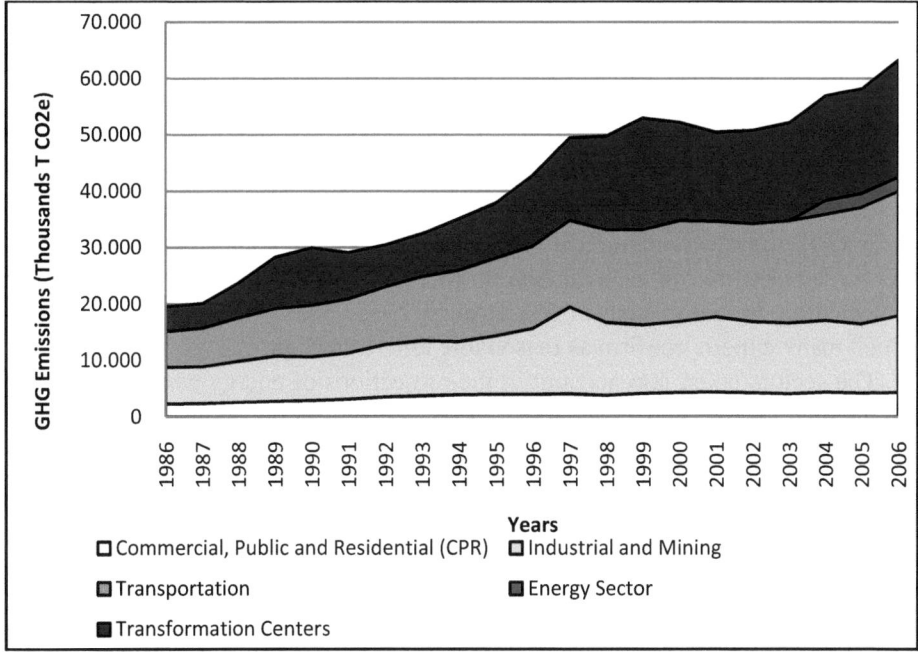

Source: developed in-house

In terms of all emissions, these rates doubled in 2006 relative to 1990 and have increased 2.2 times relative to 1986.

III. The Baseline Scenario: Projecting Energy Consumption and Emissions

This section presents the estimated energy consumption by the major fuel consuming sectors, and then estimates the GHG emissions associated with this consumption in business as usual or baseline scenarios. These projections are a first step in defining a national mitigation strategy.

We estimate long-term, end-use consumption using an econometric approach. Once the end-use demand for the various fuels is established, especially electricity, the model used can estimate the electrical requirement needed to meet those demands, which then allows for an estimate of GHG emissions from electricity generation. The consumption of other processing plants and their emissions are estimated using another methodology.

A. Forecasting End-Use Energy Consumption and Emissions

The methodology for projecting 2007-2030 energy consumption employs an econometric approach that disaggregates the different sectors from the Energy Balance published annually by the National Energy Commission. This approach enables the capture of long-term trends for energy consumption, in the various sectors, as a function of the variables affecting energy consumption in each sector or subsector. This is an appropriate approach, given the available data.[2]

A GDP trend was required in the projection as an explanatory variable. For this, in the baseline we use real data up to 2006, and for the following years, we projected an annual growth of 5% through 2015, and 4% annually until 2030, which many experts confirm as reasonable long-term figures.

The sectors taken into account in the projections of energy consumption are: commercial, public and residential (CPR), transportation, industry and mining and energy industry consumption. Because each sector is composed of subsectors with different energy consumption structures, depending on the case, the most important subsectors are broken down to estimate their consumption.

In particular, both CPR and energy are treated as major sectors. The transportation sector is broken down into: land, air, rail and maritime. Close to 70% of the energy consumption within the transportation sector is linked to land transport, while 23% is linked to maritime transport. The other subsectors account for a much smaller proportion of consumption.

The industrial and mining sector is broken down into the following subsets: copper, paper and pulp, miscellaneous mines and industries, cement and others, with the first three subsectors accounting for 90% of the total consumption for the sector. The "others" sector is made up of various subsectors such as iron, nitrates and fishing, among others.

Projected energy consumption is summarised in the following figure.

---

[2] Because long-term trends are sought, variables which generate short-term fluctuations, such as price, are not included. Another important reason for not considering price in the model is that no reliable and credible long-term projections are available for oil or carbon prices; given the high price volatility, obtaining good estimates is very difficult even in the short run. Also, saturation effect for energy use is not included in the model (other than historical data) because, while it is appropriate for developed countries, Chile's per capita GDP is not yet at those levels.

Fig. 2: Projected share of consumption 2007–2030 by sector (teracalories)

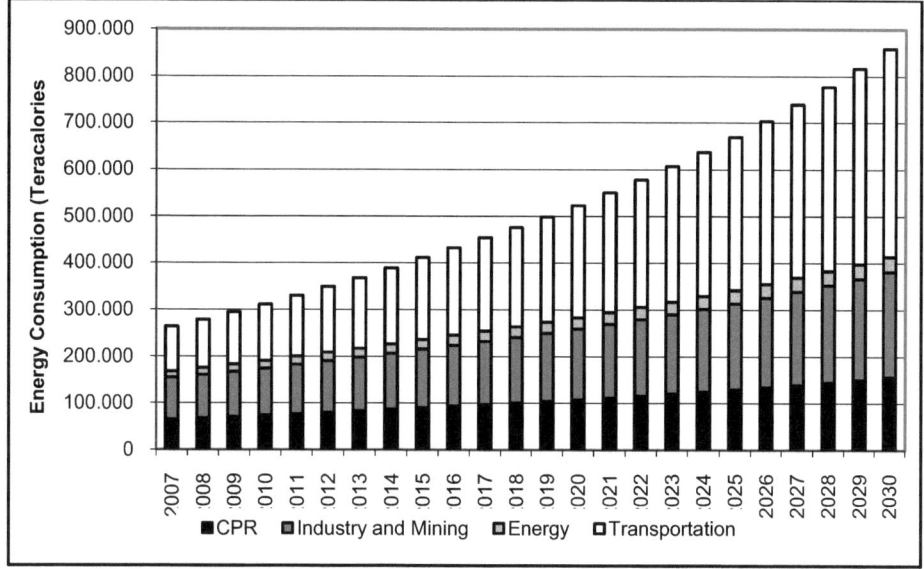

Source: developed in-house

As shown in the graph, the country's total consumption increases to 3.6 times the 2006 level (approximately 240,000 teracalories), reaching 858,000 Tcal. In 2030, the biggest sectors are projected to be transportation and industry and mining, at 52% and 26% respectively.

With regard to the energy sources most in demand in 2030, diesel, fuel oils and electricity will top the list, whereas gasoline and firewood, which were the principal sources in 2006, will have reduced usage.

The emissions estimates are obtained using the LEAP software and the IPCC's official emissions factors. The following figure shows that emissions increase 3.5 times relative to 2007, reaching a total of 186 million tons of CO_2e. Transportation is the fastest growing sector, accounting for 70% of the total by 2030, with the industrial and mining sector a distant second at 20%.

Fig. 3: Projected emissions from final demand fuel use 2007-2030

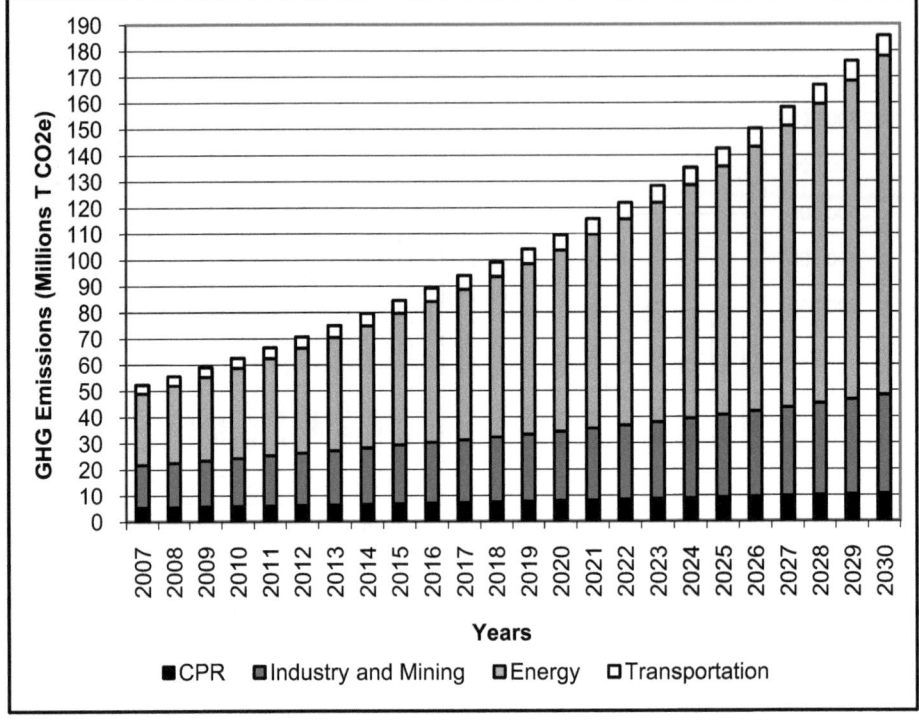

Source: developed in-house (million t of CO_2e)

B. Transformation Centre Demand Forecast

In this section, we present the assumptions considered and results obtained in the electricity generation sector and other transformation centres. Crucial assumptions in electricity generation include considering the April 2008 work plan from the Chilean National Energy Commission (CNE) to be the official forecast to 2018.[3] Besides, we do not consider the ENRC (non-conventional renewable energy) law to be part of the generation baseline scenario, as it only includes those projects covered by the work plan.

---

3 The CNE's work plan is augmented by a study carried out by the *Revista Electricidad Interamericana* (The Inter-American Electricity Magazine), which offers an analysis of future investments in Chile's electricity sector. The study is based on information from secondary sources and the SEIA (Environmental Impact Assessment System), and is considered by several experts to be fairly representative of Chile's generation plant in 2013.

Given these parameters, the existing record is modelled using the LEAP software. Installed capacity is introduced exogenously to satisfy demand and plants are dispatched by merit order. Given market characteristics and fuel prices, it is expected that the majority of the installed capacity to meet demand in 2030 will be achieved with coal.[4]

Installed capacity increases from 13,000 MW in 2007 to 40,000 MW in 2030, which is slightly more than a threefold increase. In addition, while the energy matrix in 2007 is driven by natural gas and hydropower, it will gradually transform to being 52% coal-powered. A similar situation occurs with electricity generation, which also increases threefold relative to 2007 and becomes 60% dependent on coal.

According to the methodology, emissions from electricity generation have been estimated separately from all other transformation centres. The following figure shows the projected emissions estimates from electricity generation.

Fig. 4: Projected electricity generation emissions 2007-2030

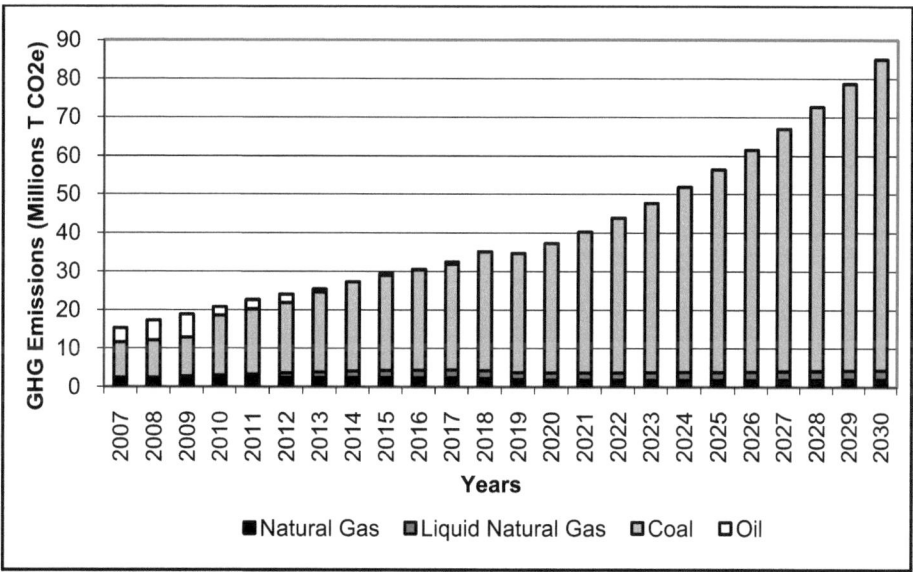

Source: developed in-house (million t of CO_2e)

The "other transformation centres" category corresponds to consumption for the processing of gas and coke (gas power and steel), oil, natural gas and methanol. These

---

4   This is a noteworthy assumption given the difficulty in evaluating the possible impact of other competitive technologies (hydroelectricity in particular). In any event, even if some of the installed and generating capacities change, this will not significantly modify the analysis and trends presented.

inputs are mainly coal, natural gas, coke and tar. In 2006, consumption amounted to 38,000 Tcal according to CNE. The methodology used in the consumption projections is an econometric regression analysis, which essentially responds to changes in GDP. The results show that total energy usage reaches 95,000 Tcal in 2030, with natural gas being the primary energy source, accounting for 86% of capacity.

Including emissions from the other transformation centres in the aggregate total does not add much to the analysis, given the relative size of the electricity sector compared to the other transformation centres. In effect, the electricity sector is responsible for 70% of GHG emissions from the transformation centre sector in 2007, and its relative importance increases through 2030, at which time it will account for 83% of the sector's emissions. In total, the emissions from the transformation centre sector increase 4.6 times relative to 2007, reaching 102 million t CO_2e in 2030.

C. Chile's net GHG Emissions

Fig. 5: Energy sector projected emissions 2007–2030

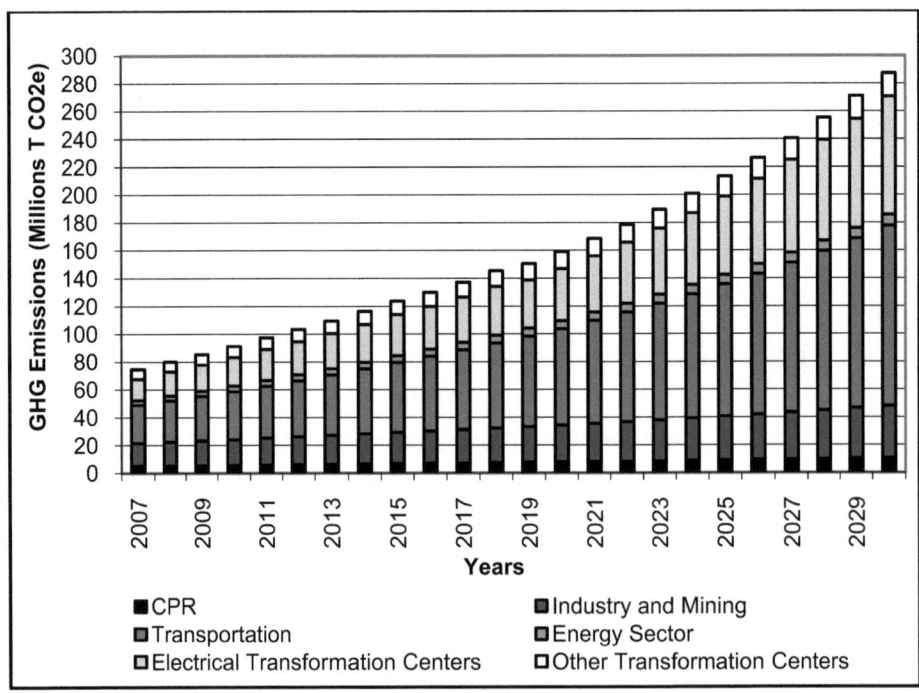

Source: developed in-house (million t CO_2e)

The final result for energy sector GHG emissions is obtained by consolidating the emissions from final demand and transformation centres. The final 287 million t CO_2e figure is 3.9 times greater than 2007, 7.9 times greater than 1994 emissions, and 9.6 times greater than the 1990 value.

The two main emitting sectors are transportation and electricity generation (including own consumption). The first sector increases its proportion from 37% in 2007 to 45% in 2030, while the second sector increases from 20% in 2007 to 30% in 2030.

For this period, two indicators of interest are the emissions per capita, and per GDP. In the first case, the indicator increases from 3.6 t CO_2/inhabitant in 2005 to 13.8 t CO_2/inhabitant in 2030. This is not a good result, as this number is higher than that of all reference countries and regions in Table 1[5] in 2005, except the United States (19.6 t CO_2/inhabitant).

With regard to GDP, the indicator moves from 0.6 kg CO_2/US$2,000 in 2005 to 1.0kg CO_2/US$2,000 in 2030. Once again, this is not a good figure as it is higher than that of all the countries and regions in Table 1 in 2005, except for China and the non-OECD European countries (with 2.7 kg CO_2/US$2,000 and 1.7 kg CO_2/US$2,000 respectively).

Table 1: International GHG emissions indicators (2005)

Country/region year 2005	Emissions per capita (t CO_2e)	Emissions/GDP (t CO_2e/ US$2000)
United States	19.6	0.5
Colombia	1.3	0.6
China	3.9	2.7
Latin America	2.1	0.6
Non-OECD Europe	4.9	1.7
OECD Europe	7.6	0.4

Source: International Energy Agency, 2005 and in-house preparation

Furthermore, if one considers that Europe intends to reduce its levels by 20% relative to 1990, and by 30% if an international agreement is reached, the Chilean levels over the next 20 years will further widen the gap between Chile and the developed world.

---

5 This table presents indicators of per capita emissions and emissions per GDP of other countries and regions that have served as reference for the analysis presented.

IV. Abatement Potential and Feasibly Implemented Economic and Regulatory Mechanisms

Chile's potential to reduce GHG emissions is dependent on a series of factors, including the productive sector being analysed, the intensity of the energy consumption (related to energy efficiency) within the sector, and the level of technological development.

This paper focuses on improvements in energy efficiency, technological developments and management capacity applicable to the various GHG emitting sectors. The measures considered are those which make sense for countries with levels of economic development similar to Chile's, where it is not realistic to assume that efficient but costly technologies which reduce economic competitiveness, or diminish the well-being of consumers, will be implemented.

Bearing these limitations in mind, this section analyses the most cost effective reduction options for the transportation, industrial, mining, commercial, public and residential, and electricity-generating sectors. The measures are presented according to sector, cost per ton of CO_2e reduced, and overall reduction potential through 2030.

The methodology used to develop this analysis considers the following aspects: international literature review, in particular studies that have already evaluated the reduction potential for a series of measures in different productive sectors; direct consultation with Chilean experts[6] regarding costs, emissions levels, feasibility, applicability, acceptance and adaptation of the various technologies, etc.; and establishment of the cost of the measure, including its underlying costs and a forecast for its potential penetration. The cost of the measures corresponds to the real cost to society of adopting them. It accounts for investment, operation and maintenance costs, and includes the savings as input or fuels as they occur. It does not take into consideration some of the social costs or benefits such as the increase in time or the inconvenience associated with the need to use public transportation if the cost of parking increases. Qualitative identification of other possible impacts (positive or negative) of the measures which could favour or impede implementation. Calibration of the technological costs to those obtained from official organisations and international literature.

Selection and evaluation criteria for the measures have been established in order to include only those which are feasible for use in Chile over the study's timeline. While many studies only include measures with costs below $50 per ton of reduced CO_2e, this study includes more expensive measures if they are feasible for implementation in Chile.

---

6 The experts are professionals from the Departments of Electrical Engineering, Mechanical Engineering and the Program for Energy Study and Research (PRIEN) at the University of Chile, Chile Foundation (Fundación Chile), and the National Energy Commission (CNE), among others.

The groups of measures considered, ranked from those with the least impact to the most impact in terms of emissions reduction potential, are as follows:

- Demand management, technological improvement and reduced transportation activity
- Demand management in the commercial, public and residential sector
- Demand management in the industrial and mining sector
- Reduction of carbon intensity in electricity-generating technologies

Based on the costs and anticipated reductions from the measures evaluated, it is possible to determine the country's abatement potential by level of reduction cost per ton of CO_2e, and a feasibly implemented abatement curve. The abatement curve below combines the evaluations of costs and reduction potentials of the measures discussed. The curve presents the abatement costs (height of the bars) and the reduction potential (width of the bars) for the main measures, namely those with the greatest reduction potential.

Fig. 6: CO_2 abatement curve for Chile

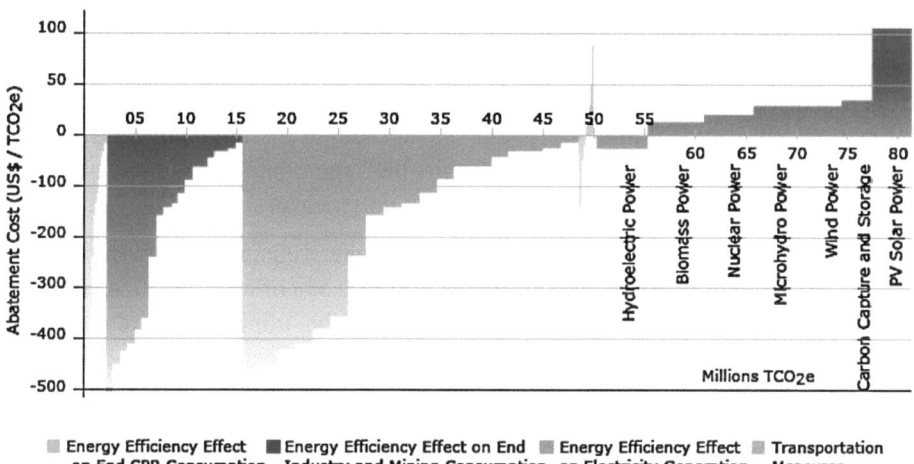

Source: developed in-house

As can be seen, there are a large number of measures with negative costs, indicating a net benefit or savings for the economy over the measures' life cycle. The measures with positive values indicate that the option has incremental costs over its life cycle relative to the reference case.

The negative cost measures primarily rely on exploiting the high potential for energy efficiency available within the country across different economic sectors, including transportation, industry and mining, commercial, and public and residential. In the majority of these cases, implementation costs for these measures are very low (promotion, regulation, certification, minor technological improvements, etc.), while the benefits measured in fuel savings are very high.

In terms of reduction potential, energy efficiency measures are most effective in the reduction of greenhouse gases, so there is a wide range of potential reductions available which have either negative or low costs. The figure above shows several measures related to non-conventional renewable energy (NCRE), as well as electricity-generating technologies with no direct emissions of greenhouse gases, like nuclear energy. These technologies exhibit positive costs.

V. Analysis of GHG Emission Reduction Scenarios in Chile

The following section presents the results from an "early actions" scenario (which incorporates the initial measures that have already helped to reduce GHG emissions but were not considered in the baseline) along with a maximum reduction scenario that includes the maximum number of potentially combined measures.

Two types of aggregate scenarios have been defined for assessment. The first, early action scenario considers the impact of reductions from the NCRE law which went into effect in 2009, along with the reductions achieved with the Transantiago project, which is undergoing improvements en route to final implementation.

The second scenario corresponds to the aggregate reduction potential achieved by joint implementation of the following measures:

- More stringent NCRE. Corresponds to a higher requirement for renewable energy generation levels. 1% of total (up from 0.5% of total) in 2014.
- Nuclear/hydroelectric power. Corresponds to the implementation of 1,000 MW of either nuclear or hydroelectric generating capacity instead of diesel or coal technology (depending on cost) starting in 2025.
- Carbon capture and storage penetration. Corresponds to the implementation of CCS technology in two coal-fired power plants with a total of 500 MW, starting in 2025.
- Specific transportation measures.
- Energy efficiency measures. Corresponds to the assessment of energy efficiency measures proposed in the National Energy Efficiency Program (PPEE). They include the industrial and mining and the commercial public and residential sectors. The energy efficiency goal is quite ambitious and, for this reason, this scenario is quite demanding in terms of reduction in energy consumption

This estimate is developed using the LEAP model which allows for complex analysis beyond a simple aggregation of the potential of individual measures. The model can calculate the interaction effects from the various options' effects on sector energy demand, altering the consumption variables (and emission variables) both in calculating end demand and transformation centre demand. This analysis is shown in the following figure.

The potential reduction in greenhouse gases is 80 million tons of CO_2e (equivalent to a 28% reduction relative to total projected emissions in 2030). Most of these reductions relate to energy efficiency measures in the industrial and mining and commercial, public and residential (CPR) sectors. Additional significant reductions (direct and indirect) come from the electricity transformation sector.

This abatement potential reflects the combined impact from both the economic and technological assumptions as well as the policies necessary to achieve their implementation. The previous section analysed the regulatory and economic instruments needed to promote the effective penetration of these measures in the specified sectors, within the defined time horizon.

Fig. 7: Maximum GHG emission reduction in Chile

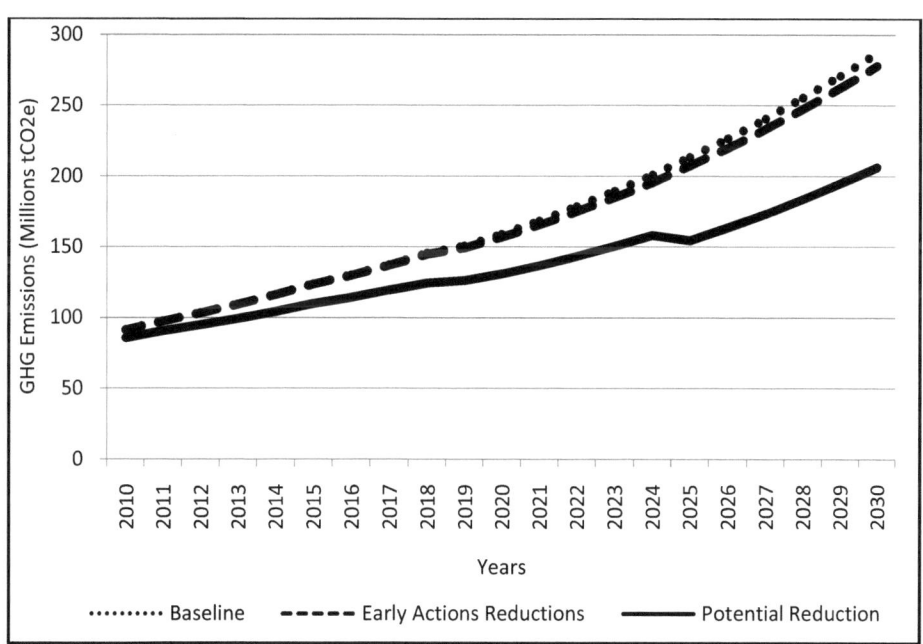

Source: developed in-house

The final result for energy sector GHG emissions is obtained by consolidating the emissions from final demand and transformation centres. The final 287 million t CO_2e figure is 3.9 times greater than 2007, 7.9 times greater than 1994 emissions, and 9.6 times greater than the 1990 value.

The final result for energy sector GHG emissions is obtained by consolidating the emissions from final demand and transformation centres. The final 287 million t CO_2e figure is 3.9 times greater than 2007, 7.9 times greater than 1994 emissions, and 9.6 times greater than the 1990 value.

Some measures, such as energy efficiency in the industrial and mining and CPR sectors, require low economic investment to be implemented, but do require strong incentives and promotion by both the government and the industries themselves. On the other hand, options such as implementing a higher proportion of electricity generation through NCRE or CCS will require considerable injection of funds, in addition to a push from government and industry, and even consolidation of these technologies through their application on a commercial scale.

The following figure shows each sector's abatement potential and relative importance in 2030. Additionally, the subsequent table shows the absolute values in each case.

In the event that the maximum potential is achieved in 2030, the primary responsible GHG emitting sector would be transportation, which increases from 45% in the baseline scenario to 63%, followed by industry and mining, which would reduce its participation to 11%. Electricity transformation centres decrease from 29% to 9%.

With respect to the total cost of implementing these measures, a separate analysis must be made for the electricity generation sector. When it comes to the CPR, industry and mining, and transportation sectors, the cost of implementing the measures is negative (a net benefit is obtained). Thus, a high percentage of the maximum potential GHG emission reduction can be achieved from these sectors without incurring any direct costs.

The final result for energy sector GHG emissions is obtained by consolidating the emissions from final demand and transformation centres. The final 287 million t CO_2e figure is 3.9 times greater than 2007, 7.9 times greater than 1994 emissions, and 9.6 times greater than the 1990 value.

The final result for energy sector GHG emissions is obtained by consolidating the emissions from final demand and transformation centres. The final 287 million t CO_2e figure is 3.9 times greater than 2007, 7.9 times greater than 1994 emissions, and 9.6 times greater than the 1990 value.

Fig. 8: Sector potential of GHG reduction and relative importance in Chile 2030

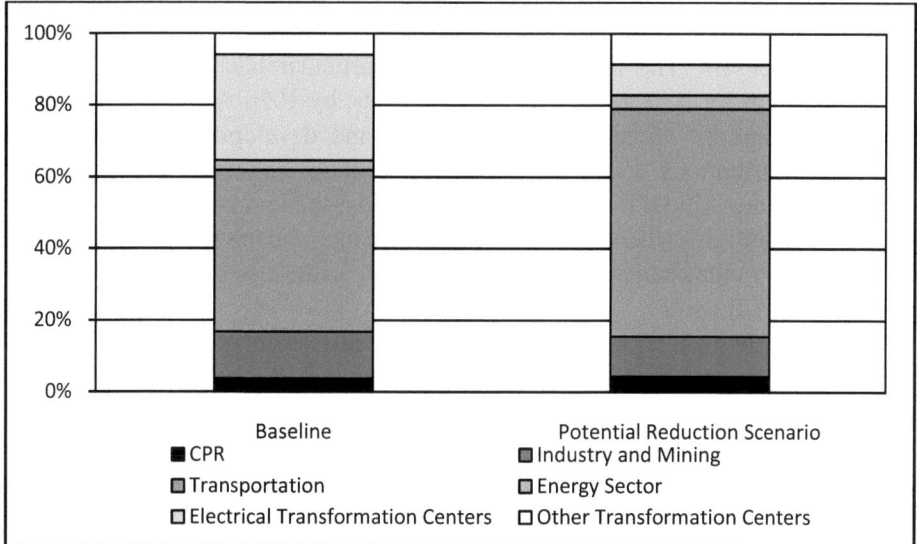

Source: developed in-house

The same is not true for the electricity-generating sector. If all of the CCS, a more stringent NCRE law, and an increase in hydroelectric and nuclear generation options were implemented, seeking the maximum possible reduction, the total cost would reach a little over US$1 billion a year (relative to the reductions expected in 2030).

VI. Conclusions and a Basis for a Chilean Greenhouse Gas Mitigation Strategy

Global climate change and the resulting need for international policies and requirements have become major topics on the agenda of developed countries. The available scientific data make it clear that the time for discussion has passed, and short-term actions are necessary to mitigate emissions, if climate change is to be kept within reasonable margins. It is also evident that both developed and developing countries will have to contribute significantly to the reduction of greenhouse gas emissions.

Negotiations on the framework that will govern the required GHG reductions are being carried out by various groups of countries, as well as some significant individual countries. While they advocate many shared positions, there is also difference of opinion about the future requirements for GHG mitigation. Of the industrialised countries, Canada, the European Union and Japan have decided to work towards reducing global emissions by at least half by 2050, albeit with different proposals. The European Union in particular has pledged to reduce GHG emissions by 20%, compared to 1990 levels, by 2020, and by 30% if other developed countries commit to similar levels and developing countries make adequate contributions. Japan favours a proportional, sector-based approach for future reductions. Under the current administration, the United States is expected to take more initiative when it comes to climate change, and has signalled interest in resolving their emission situation domestically, while essentially pushing for technological solutions.

The only developing countries with significant weight in these discussions are the G5 countries: Brazil, China, India, Mexico and South Africa. Their influence stems primarily from their role as significant emitters and/or the size of their population. The G77, which includes the majority of developing and Latin American countries, both emerging and very impoverished, as well as oil-exporting countries, comprises a very heterogeneous grouping of countries. Its effectiveness, in particular to advocate for positions relevant to Latin America, has been declining in recent years.

The negotiation process and positions taken have three areas of concern for Chile, and Latin America in general. The first is the discussion of the strong need for strategies to control the increase in emissions by developing countries. This is a reality that Chile must confront responsibly. However, in these discussions, no fundamental distinctions are made that could account for the real situation in most Latin American countries, which is very different from those in China, India or Brazil.

Secondly, the basic background information necessary for countries in Latin America to define a position on mitigation is scarce and/or of dubious quality, particularly in comparison to the wealth of information, studies and resources devoted to this area in developed countries. It is difficult to establish and defend a position based on available information.

Finally, Latin America (aside from Brazil and Mexico) is not represented in the appropriate forums – either individually or as a group – to make and defend commitments that are realistic for the region. As a result, Latin American countries end up joining agreements made by a large group of countries which differ widely in terms of technology and productivity levels, expectations for future development, and GHG emissions. This is different from what the case was in the early nineties when Latin America was a protagonist in making proposals for moving forward with commitments that balanced the need for economic growth and a responsible approach to climate change.

The results of this process may complicate the future development of Latin American countries insofar as not joining in the agreements could mean difficulties for our exports, and difficulties as well in gaining access to credit and investment. On the other hand, joining in these agreements might also come at a very high cost to the industry and inhabitants. The main problem is that controlling emissions can be more costly for many Latin American countries than other developing countries because the energy or productive sectors are much higher emitters and less technologically advanced. In other developing countries, limiting emissions growth could be relatively simple to achieve through new technology which also leads to greater profitability over the medium term. Furthermore, some Latin American countries have based their energy needs on relatively clean sources, particularly hydroelectric power. This will be difficult to maintain in the future as lower-cost renewable energy resources are depleted.

Given the context detailed above, the definition of nationally appropriate mitigation options must be undertaken immediately. The results presented in the preceding sections offer sufficient quantitative data to define a realistic Chilean base strategy for greenhouse gas mitigation. Some specific recommendations for a future strategy are made in the following section using this data. Future studies must delve into additional topics such as emissions from the agricultural sector, the potential effectiveness of carbon capture associated with land use change, and expected reductions from energy efficiency policies and support for development of clean energy sources.

The following conclusions have been drawn and are relevant for defining a framework within which to develop a mitigation strategy proposal for Chile.

Baseline

- Chile is and will be a country whose GHG emissions will be largely irrelevant on a global scale. This is also true for Latin America as a whole (excluding Brazil and Mexico).
- Unitary GHG emissions indicators for Chile will significantly worsen according to the baseline scenario through 2030.
- Energy-related and GHG processes emissions in Chile have doubled over the last 16 years (1990-2006).
- In the next 16 years (2007-2024), energy-related and GHG processes emissions will increase 2.9 times. In 2030, they will be 4.2 times greater than in 2007.
- The sectors that will increase emissions the most through 2030 are electricity generation and transportation.
- Over the short term, the electricity sector will continue to be the second largest emitter of GHG in Chile.

- Carbon capture due to land use changes (LUC) is on a decreasing trend through 2030.

Mitigation options

- Early actions by Chile would reduce energy and process- related emissions by 3% in terms of projected 2030 emissions.
- Energy demand management measures will be key in achieving significant GHG emission reductions with respect to the baseline.
- Aggressive introduction of NCRE will allow for major reductions in GHG emissions.
- The incorporation of a 1,000 MW nuclear power plant in 2025 and increased hydraulic capacity on the same scale, beyond what is included in the work plan, would add a 3.3% emission reduction through 2030.
- Assuming that 2 or 3 coal-fired power plants (1,000 MW) apply carbon capture and storage (CCS) starting in 2025, the estimated potential emission reduction would be 1.4%.
- The measures identified for the transportation sector only marginally reduce sector emissions.
- All the measures described above would allow for a total reduction of 28% compared to the expected 2030 emissions.
- Even if all identified measures are applied, total energy and process-related emissions in Chile will more than triple between 2007 and 2030.
- An active international carbon bond market with relatively high prices per ton of CO_2e reduced in generation would allow for an additional 9% reduction in emissions by 2030.

Costs

- Cost per ton of CO_2 reduced is highly variable.
- Energy efficiency, of the principal reduction measures available, provides net benefits to those who use them.
- Nuclear and hydropower plants would allow reductions at relatively low costs.
- The incorporation of non-conventional renewable energies, with the exception of biofuels, has a relatively high and variable cost.
- CCS has an estimated cost of between US$65 and US$80 per ton of CO_2 reduced.
- Although measures oriented towards reducing automobile traffic and improving truck efficiency do generate important net benefits, they do not generate significant reductions.

Recommendations for a Chilean Strategy

Taking the findings thus far into consideration, the proposal of a set of general recommendations regarding an appropriate mitigation strategy for Chile is as follows:

- The principal recommendation from this study is that, while Chile must make commitments to mitigation, these commitments must not compromise Chile's future development which will require an increase in GHG emissions. This means that Chile must strongly defend the principle of shared but *differentiated* responsibilities.
- Chile must define and defend a projected baseline as the benchmark against which to measure its performance.
- Chile needs to actively participate in the proposal and definition of performance indicators used to measure Chile's commitment.
- Chile needs to proactively develop a plan of international alliances within Latin America.
- Chile must design and defend an inter-sectorial "climate-friendly Chile" strategy.
- In the various forums, one must emphasise that hydropower development is an important part of a climate-friendly strategy. The key is to begin working, in an integrated manner, with the affected communities on the development of projects which will minimise the local impacts that are currently hindering their development.
- There should be wider use of international instruments which encourage viable technological changes in the energy matrix.

Despite the facts listed above, which are focused on an external strategy and Chile's participation in international forums, Chile also needs to consider the following four domestic actions in developing its strategy:

a) Stimulate coordinated actions through institutional change and the development of tools to achieve significant reductions of GHG emissions in multiple sectors, while allowing Chile to continue growing economically and improving the quality of life for its inhabitants.

This will require:

- Development of an institutional framework that is a top priority on Chile's political agenda.
- Financial institutions which promote emission reduction projects.

- Provide visible and consistent national-scale signals over time to enable appropriate investment decisions
- Highlight the opportunities associated with greater energy efficiency.
- Create specific initiatives to overcome the economic barriers to the introduction of NCREs and greater energy efficiency.

b) *Quickly and strongly support measures geared towards greater energy efficiency and the implementation of negative cost measures*

Specifically:
- Demonstration projects need to be developed in the transportation sector.
- International support is required for measures which have a greater impact on transportation
- The potential from better demand management is significant enough to justify strong and active development of options for various Chilean sectors, particularly if they reduce electricity consumption.

c) *Begin examining low-carbon generation options now*

It is also time to begin to:
- Carefully study the options for incorporating a nuclear plant within the next 15 to 20 years
- Conduct research on the applicability of wind, geothermal and solar energy on a massive scale
- Explore the potential for carbon capture and carbon capture and storage (CCS)
- Study the possibility of increasing storage capacity for liquefied natural gas in Chile

d) *Promote a Latin American level plan of international alliances to defend the interests of the countries in the region.*

This includes considering that:
- Chile cannot simply follow its "own path" with respect to its international obligations.
- Chile, and Latin America in general, (excluding Mexico and Brazil) can take advantage of low emission weights on a global scale to press for a scheme appropriate to their future development.
- Public/private alliances are required to promote the creation of a national strategy.

VII. References

[1] M. Bohm, H. Herzog, J. Parsons and R. Sekar (2007). "Capture-ready coal plants – Options, technologies and economics", International Journal of Greenhouse Gas Control 1, pp. 113-120.
[2] C. Böhringer and A. Löschel (2005). "Climate Policy Beyond Kyoto: Quo Vadis? A Computable General Equilibrium Analysis Based on Expert Judgments" KYKLOS, 58(4), pp. 467-493.
[3] Centre for Technology Strategy (2007). EJTIR, 7, no. 1, pp. 15-38.
[4] Climate Change (2006). Assessment of technologies for CO_2 capture and storage, Climate Change 2006.
[5] Federal Environment Agency (2007). A Climate Protection Strategy for Germany – 40% Reduction of CO_2 Emissions by 2020 Compared to 1990, http://www.umweltbundesamt.de, Germany, August 2007.
[6] GTZ (2003). "National Strategy Study (NSS) for the CDM in Chile". Summary in Spanish. Climate Protection Programme (CaPP), July 2003.
[7] IEA/OECED (2005). Projected Costs of Generating Electricity, 2005 Update. International Energy Agency, Organisation for Economic Co-Operation and Development.
[8] IEA/OECED (2007). Tackling Investment Challenges in Power Generation in IEA countries, International Energy Agency, Organisation for Economic Co-Operation and Development.
[9] McKinsey & Company (2007). Reducing U.S. Greenhouse Gas Emissions: How Much at What Cost?, December, 2007.
[10] MIDEPLAN (2008). "Social Prices for Evaluating Social Projects". Ministry of Planning.
[11] MTT (2005). "Analysis of Energy Efficiency in Interurban Cargo Transport". Conducted by: CIMA Ingeniería EIRL for the Undersecretary of Transportation.
[12] Raúl O'Ryan, M. Bosch and E. Matamala (2002). "Estimate of the Impact of Transportation Measures on Emissions from Mobile Sources in Santiago". Ingeniería de Sistemas, vol. XVI, no. 1, 93-119, Department of Engineering, School of Physical Sciences and Mathematics, U. of Chile, June 2002.
[13] Progea (2008). "Comparative Analysis of Light and Medium Vehicle Emissions: Policy Proposals". Program for Environmental Economics and Management of the University of Chile. Commissioned by the National Automotive Association (ANAC). May 2008.

[14] Progea 2 (2008). "Economic Analysis of the National Energy Efficiency Program's 2007-2015 Strategic Plan for Chile". Requested by Subsecretary of Economy, Development and Reconstruction.
[15] P. Reinelt and D. Keith (2007). "Carbon Capture Retrofits and the Cost of Regulatory Uncertainty". The Energy Journal 28(4), pp. 101-127.
[16] U. Springer (2003). "The market for tradable GHG permits under the Kyoto Protocol: a survey of model studies". Energy Economics 25, pp. 527-551.
[17] L. Viguier, H. Babiker and J. Reilly (2003). "The costs of the Kyoto Protocol in the European Union". Energy Policy 31, pp. 459-481.

A Methodological Proposal for Community Participation in the Development of Microgrid Projects

N. Garrido, M. Álvarez, G. Jiménez-Estévez

Abstract

The Cóndor Sustainable Electrification Project (ESUSCON) was established with the goal of optimizing the use of local renewable energy resources, together with conventional systems, by coordinating the operation of those small electricity generators on a small scale and within a common geographical location. This work promotes technologies used to compete in terms of land use, and the renewable natural resources within the framework of rural space use, which might give rise to various types of conflicts, problems, or claims on the part of the participating communities. To minimize or avoid these, the community involved drew up an important role during the implementation of the project, which made the integration of the community fundamental in the decision-making process during the construction, implementation and operation of the system. To facilitate such openness, it is important to have previous knowledge of the form of organization and relationships, valuation of the immediate surroundings, needs, etc. In the case of the Huatacondo locality, a method of participation was implemented in that an agreement with the characteristics of the community permitted generating greater empowerment and, at the same time, sustainability of the project. This was obtained by taking into consideration the facts that the participating communities are well-known, their territory is appreciated and they should coexist with projects of this type.

I. Introduction

During the past few years, interest in social acceptance of energy projects has increased. From a macro point of view, these projects have matured considerably, but it is still too complex from the local point of view. One of the reasons for this is that although these new energy technologies are interesting from a collective perspective – mainly due to the reduction of greenhouse gas emissions or a decrease in fossil fuels consumption – local projects must, on the other hand, deal with different interests of the community[1]. That is why in the first instance it is important to identify that the project is not under negotiation with politicians, residents, social organizations or other local representatives[1]. In this way, it will be possible to generate a permanent dialogue with the different players involved, generating confidence between the community and the professionals in charge of the project[2].

The Cóndor Sustainable Electrification Project (ESUSCON) was established by the Energy Center of the Department of Electrical Engineering in the School of Physical Sciences and Mathematics at the University of Chile in the village of Huatacondo, located in the town of Pozo Almonte in the Tarapacá region. This project was developed under the concept of a virtual generation, which tries to coordinate the operation of electricity-generating unit groups, located in a common geographical zone on a small scale, and which takes advantage of the local renewable energy resources, achieving a system operating like a single conventional generator[3].

ESUSCON and the associated technologies of renewable energy compete for the use of land and natural renewable resources within the standard framework for using the participating rural space in a way that, without adequate participation, could generate problems and/or environmental conflicts. Concomitantly, the community could present polarization and resistance to the project (if the population feels excluded during the decision-making process since the project directly affects the geographical area and quality of life of the community)[4]. Furthermore, it should be emphasized that these types of technological interventions bring about social alterations because of the planned changes in the culture of the participating community[5]. This planned change must respect and recognize the life standard parameters that the local community desires and also its relationships and power structures[6]. That is why it is also necessary to define the type of sustainable development desired by the intervened community, which must then be based on the local interpretation and decided upon jointly with them[7]. If the previous consideration is taken into account for a technological intervention, social empowerment of the project may be achieved.

As a result of the issues presented above, the Social Introduction Team sought a proposal that permitted adequate participation and integration of the Huatacondo community; that included sustainable development and that minimized or avoided conflicts associated with the implementation and acceptance of the project. To accomplish this aim, they proposed the following objectives:

- Determine the feasibility of implementing the project in the community being studied.

- Contribute an implementation and development plan within a framework of sustainability, through a proposal of criteria and recommendations derived from the opinions of the community involved.
- Obtain empowerment of the community in the project through participatory processes that involve the inhabitants in the development of the project in its long-term management and operation.
- Conduct an environmental evaluation that includes environmental, social and economic dimensions.
- Accomplish monitoring and evaluation systems that permit identification of socio-cultural changes and the sense of belonging on the part of the community.

II. Methodological Proposal

The methodology has been divided into 4 stages that facilitate participation like that accomplished in the Huatacondo locality, in a way that avoids opposition and negative effects, both for the development of the project and for the community.

A. Stage I. Identification of the Participants and Characterization of the Village

Concerning community work, it is quite important to build confidence among the parties involved, i.e. the local community and professionals in charge. Achieving this confidence requires time and patience; given that it is necessary to create spaces for dialogue that allow growth in confidence. These spaces can be: meetings in households, sharing food with some inhabitants, field trips with residents, meetings with small groups, information dissemination, and workshops among others things[2].

Stage 1 consists of an initial approach with the community for the purpose of accomplishing a social and territorial characterization of the village, identification of the relevant local participants and knowledge of the disposition of the population who participate in the decision-making process[2]. For this, an updated bibliography was obtained to allow identification of public and private institutions that have a relationship with and knowledge of Huatacondo at the national, regional and community levels.

At the same time, the most relevant variables of the locality are defined and the energy situation determined: i.e. the habits of the consumers, expectations, needs and forms of energy use. This follows on from secondary information and site visits that present two scopes of action: direct observation and semi-structured interviews. These are divided according to the dimensions of environmental sustainability, i.e. social, economic, environmental and institutional. In this way, all the consultations required for the participatory development of the project are realized.

B. Stage II. Integration of the Community in the Project

Once the social and territorial characterizations are accomplished, the team has a meeting with the community to reveal the results and explain the analysis made. Then the participants perform a social validation that permits corroboration and augmentation of the information compiled in the first stage, giving clarity to the perceptions of the community[8].

After the social validation, it is possible to commence with the direct work of integrating the community with the project, now that there is clarity about the form of organization of the village, its interests, needs, and so forth.

The work with the community was divided into two types of activities: classes and participatory workshops for adults and children.

Table I: Informational and participatory activities for adults and children

Classes		Workshops	
Adults	**Children**	**Adults**	**Children**
Explanation of the Project	Renewable Energies	Social Validation	Definition of Zones of Location
Renewable Energies	Energy Demand Management	Definition of Zones of Location	Social Cartography
Energy Demand Management		Social Cartography	Map of Wishes
		Social Map	Choosing a Name for the Project
		Map of Wishes	
		Choosing a Name for the Project	

The general method used for the workshops consists of gathering the whole community in one place, explaining what the activity consists of, welcoming comments, suggestions and ideas, and finally, discussing the results together. Once the workshops have ended, those people who did not attend are informed, door-to-door, of the decisions made by those who attended, and are consulted about whether they agree with them.

C. Stage III. Environmental Analysis and Evaluation

The environmental evaluation consists of the following phases:

a) Definition of the area of influence and elaboration of the baseline. First, the location zones of the generation units are defined, as well as those sectors that the community uses for the development of its activities. Secondly, an updated bibliography and list of consulting organizations at national, regional, community and local levels are made. Then site visits are undertaken to complement and corroborate the information obtained previously. These visits have two forms: direct observation and surveys.
b) Community perception. By means of surveys, community perception about the project and the impacts could be obtained. This information is complemented by the results of the workshops from the previous stage.
c) Evaluation of impacts. Environmental elements and components are identified that could be seen as affected by the project. At the same time, actions and activities are determined that might provoke impacts. Then a matrix of potential impacts is generated, described, valuated and prioritized.

D. Follow-up and Evaluation

Most of the sustainability indicators are performed using a top-down methodology, where sustainability is estimated from national data[9]. Therefore, critical issues related to sustainable development at local level may be ignored; hence important items for the community could not be measured[10]. Community participation in data gathering to elaborate indicators is too low, unless the evaluation is linked to clear and immediate benefits[11]. Due to this last reason, it is accepted that communities should participate in all the planning and implementation stages, including the selection, collection and monitoring of indicators[10]. There are two methodological paradigms regarding sustainability: bottom up and top down[12]. In this case, the first method is applied, considering that the community is involved in the construction of the indicators, generating joint learning between the community and professionals. Developers that follow up this working line state that it is fundamental to involve the community in the research process in order to obtain relevant perspectives about local conflicts, which also stimulates social action[13].

Once the project is implemented, the identification of positive and negative effects on the mid and long-term quality of life of the Huatacondinos is attempted. At the same time, an attempt is made to determine the sustainability of the project.

To accomplish this, indicators of sustainability are elaborated upon using variables that could affect the project and, at the same time, those that are relevant for the quality of life of the community.

Once the variables are identified, a baseline of sustainability indicators is developed which permits comparison of the parameters of the present situation (without implementation of the project) with the projected future situation.

It is also important to have periodic visits to welcome suggestions, comments, and ideas that arise from the community once the project is implemented. This enables improving details of the system that could not be identified earlier.

III. Results

A. Stage1. Identification of the Participants and Characterization of the Village

The characterizations done included social, environmental, economic and land use aspects of the Huatacondo village, which were compiled principally through semi-structured surveys in visits to 24 out of the 27 dwellings that were inhabited as of December 2009. In addition, this characterization had been previously guided by the secondary information compiled and complemented with information obtained through observations of the countryside of the Huatacondo community and its surroundings.

It was found, through the characterization process, that Huatacondo is populated by approximately 78 permanent residents, and that the majority of these people are senior citizens. About 70% of the existing houses in the village were unoccupied, used only as second homes, i.e. for vacations, national holidays and other important dates such as the celebration of the Assumption Day of the of the Virgin on August 15th. Huatacondo can be described as part of a minority of villages in the interior of the Tarapacá region that do not belong to the Aymara culture, despite the fact that the majority of its inhabitants are descended from Spaniards.

The social organization that best matches that of the village is a "group of neighbours", but there are also other social organizations recognized by the people, such as the "Committee for Light", "Committee for Drinking Water", "Committee for Irrigation", "Mothers' Centre", the school, and the church. This village does not have electricity 24 hours a day; only a schedule of 10 hours daily from Monday to Friday (2pm to midnight) and 8 hours a day during the weekends (4pm to midnight). Electricity is generated by an oil-fed diesel motor donated by the city of Pozo Almonte, and 2 mining companies near the village. Huatacondo does not have good communication with the rest of the zones, as it does not have cellular coverage and the access roads are deficient which causes the village to remain isolated (a phenomenon they call "Bolivian winter"). The Internet (satellite) only functions in a limited area around the village school, and there is only one landline telephone that belongs to a private individual.

Subsistence farming activities are carried out in the village, including the cultivation of alfalfa and corn; in the plantations, Christmas pear trees, oranges and lemons; in animal farming, pigs, goats, rabbits and guinea pigs. There is a distinct presence of condors in the vicinity of the village, after whom Huatacondo takes its name (nest of condors). Existing amongst the village's natural resources, with potential for obtaining power, are: solar, wind and biomass. In addition, the village has great tourism potential because of its petro glyphs, geo glyphs, a salt waterfall, and dinosaur tracks. This potential for tourism is not exploited by the inhabitants because there is no accommodation available or anywhere which sells food.

B. Stage II. Integration of the Community in the Project

The integration stage of the community of Huatacondo into the ESUSCON project began in December 2009 by consulting personally with the resident families on how they could participate in it. The result was that approximately 95% of the consulted families were in agreement with the realization of the project. Later, there was a process of social validation in which those attending were presented with the vision of working closely with the community that the team had previously formed from the results obtained by the social and territorial characterizations. The results met with agreement in every dimension, especially in the social, environmental and economic dimensions. Once the results of each dimension were presented, there was a period of approximately 15 minutes during which those attending added their comments. They completely approved of the team's vision of the community, and they validated the information obtained during the phase of social and territorial characterization of Huatacondo. This allowed the implementation of classes and workshops for adults and children to begin.

1. Classes

Classes started with a community talk on renewable energies. Here the sources, machinery, and functioning of solar, wind, biomass, geothermal and hydraulic energies were explained using visual media. The process of obtaining electricity through each of these energies was also presented through the medium of general diagrams. Once the potentials of renewable energies were realized, the goals, objectives and implications of the ESUSCON project were explained, also as key concepts to give better comprehension of what could happen in Huatacondo. An explanation of the Energy Demand Management followed; a system which consists mainly of a visual interface that mediates colour signals, permitting modification of the population's consumer habits in favour of making efficient use of the energy brought about by the system.

This was presented using visual resources in which a drawing was shown associated with a piece of equipment used for the Energy Demand Management system and the symbolism that they must use. At the end of each class, those attending were asked for suggestions, comments, and their impressions of what was presented. Each of the attendees was also consulted on their availability for participating in the Energy Demand Management system, with registration of the names of all those willing to participate. These same classes were given to the students in Huatacondo's School, G-101.

2. Workshops

During the last week of March and the first week of April 2010, cycles of informative and participatory workshops were held in conjunction with the community of Huatacondo. They began with delivery of pamphlets to each of the families living in the village, which explained the general form of the contents and goals of each of the workshops. At the same time, signs inviting the community to participate in the workshops were displayed in strategic places in Huatacondo, among them the social club, store, first-aid station, school, wall of the house in which the only telephone of the village could be found and on the main wall of the house where the team in charge was staying. In addition, the dates and times of the workshops were disseminated through microphones and by posting messages on the information board at the social club.

The first workshop held was on social validation, which was explained in earlier paragraphs. Later, there was a workshop on defining the zones of location, in which the proposed locations of the electricity generating units – i.e. wind turbines, solar panels, and bio digesters – were identified as affecting in any way the perception the Huatacondinos have of their village. This activity was carried out using a model of the village and its surroundings, together with figures representing each of the electricity-generating units. After the workshop, the team in charge analyzed the results and decided on the final placement of the generators; those that were ratified by the people attending the workshop, and later those mentioned to the members of the community who did not attend.

The purpose of the social map workshop was to determine the existing relationships, through a graphic representation of the relationships and interactions among the existing organisations of Huatacondo[14, 1], complemented by means of a socio-gram that allowed identification of relevant people in each of the social organizations identified earlier[14, 15].

Fig. 1. Social map of the Huatacondo locality

Once the social maps and their corresponding socio-grams were finished, a social map was drawn that permitted important points in the village, natural or artificial, to be determined[16]. Adults emphasized their crops and patron saint festivals, mentioning that these have maintained the identity of the village. On the other hand, the 1st and 2nd year children drew their appreciation for the mountains, cemetery, pools, sun, houses, vegetation, diesel generator and the water tower. The children in the 4th year through to the year 8 expressed their appreciation for the vegetation and water because both are part of the history of the village.

Finally, the "map of dreams" workshop was held in which the Huatacondinos identified their vision of their community for the future under two scenarios; one without the system of sustainable energy supply, and the other with the system of sustainable energy. This permitted them to identify whether or not they would benefit from the project[16].

For the adults, the future of the village under the first scenario is a Huatacondo with vast vegetation and little water. They mentioned that the construction of the houses would be better but their population would diminish. They were categoric in saying that the patron saint festivals maintain and conserve the identity of the village. As regards the vision under the second scenario, the participants mentioned the necessity for contact through a cellular antenna and for improving Route A-85 or creating an alternative route that would be outside of the gorge. They believe that these improvements would encourage a substantial growth in tourism.

The 1st and 2nd year children drew a school with a big library and an improved infrastructure. They emphasized the vegetation, animals, and people of the village. They added supernatural aspects like chests of gold coins, genies, phantoms and extraterrestrials. The students in years 4-8, when imagining the village of the future without the project, drew an abandoned and contaminated village. However, under the scenario of Huatacondo with the project, the children dreamed of having a cellular antenna and Internet for the whole village. They thought that the population would grow, that they would have more stores, another access route, a hotel for tourists and a better infrastructure for emergencies.

While the team in charge was there, they created a box and invited the inhabitants to deposit possible names for the project in it. Once compiled, these were put to a popular vote at the end of the workshop on the map of dreams. The name chosen by the majority of the voters was "Condor Sustainable Electrification" [Electrificación Sustentable Cóndor] (ESUSCON). The goal of this activity was to contribute to the process of empowerment of the community with the project.

C. Stage IV. Environmental Analysis and Evaluation

The results of the matrix of potential impacts and later prioritizations revealed the following components of moderate negative importance and those of high negative importance during project development:

Table II: Impacts of high and moderate negative importance

Component	Impact	Hierarchical Structure	
		Construction Phase	Operational Phase
Birds	Loss of Birds		High Negative Importance
Landscape	Change in the Quality of the Landscape	Moderate Negative Importance	Moderate Negative Importance
Social Atmosphere	Increase in Traffic	Moderate Negative Importance	
	Bad odour	High Negative Importance	High Negative Importance
	Solid Residues	Moderate Negative Importance	

The first two primary impacts are related to the installation of the wind turbine. In the case of the loss of birds, it refers to the possible collision with the Vultus gryphus (Condor). This is emphasized because it is an intelligent species with good vision and curiosity. In Huatacondo it is necessary to appreciate that the condor frequently crosses their zone.

Generally, it has a flight path, but when there are different elements from those to which they are accustomed, they come closer. In this way, a wind turbine is converted to a negative factor for this species which, if it comes close to it, could die. The change in the quality of the landscape happens in the zone where the wind turbine is located, since, according to the evaluation of the landscape done by the team, this is a zone of "Partial Retention," i.e. the interventions must be visually subordinate to the characteristics of the homogenous zones[17].

The impacts of moderate and high negative importance to the social atmosphere refer to the increase in the flow of vehicles, the generation of solid residues, and bad odour. The latter is the most relevant since it is produced by the installation of organic material for storage in that place, and organic material with the bio digester expels bad odours that could cause discomfort to the resident population.

According to the evaluation carried out, the most important positive impacts are the following:

Table III: Impacts of positive importance

Component	Impact	Prioritization	
		Construction Phase	**Operational Phase**
Water Resources	Participation in the System of Drinking Water		Positive Importance
Social Atmosphere	Increase in Permanent Employment		Positive Importance
	Change in the market by the strength of offers of temporary employment	Positive Importance	
	Increase in the Demand for Services and Consumer Goods	Positive Importance	Positive Importance
	Increase in Electrical Energy		Positive Importance
	Decrease of CO_2		Positive Importance

The positive impact on water resources refers to participation in the system of drinking water installed in 2003 by the Division of Water Works [Dirección de Obras Hidráulicas (DOH)]. This system is not actually currently in use because it has problems with pressure and intermittent supply. Intervention consists of integrating the supply of drinking water with the energy system in such a way that the water tank would have water 24 hours per day. The important positive impacts in the social environment component are related to the increase in temporary employment, a product of the accomplished civil works.

For the system to operate optimally, it is important for the people of the village to perform necessary maintenance, for example, cleaning the solar panels, making the mixture for the bio digester, revising the panel of batteries, among others. This would increase permanent employment in the area.

D. Evaluation and Follow-up

The variables used for the elaboration of the baseline include those elements that the project might affect and those that are related to the quality of life of the community. These are:

a) Scholastic results: the average grade in the G-101 school 2000-2009 was 5.9 (out of 7.0). It should be noted that the number of students in this school is low, not exceeding 12 each year.
b) Cultural transformations: the main festivals that are held in the village are the May crossings (weekends in May), the Assumption of the Virgin (14-19th August) and Cuculi (a song and dance done in the early hours of the morning on 14-19th August).
c) Security: 100% of those interviewed mentioned that there is no delinquency in the village.
d) Social organizations: the "Group of Neighbours Number 8 of Huatacondo" has 52 members who actively participate, making it the organization with the most adherents. Other organizations worthy of note are the "Center of Mothers", "Enterprising Women" and "Aymará Center",
e) Migration: to date, the population of Huatacondo is 81 permanent residents which corresponds to 33 families.
f) Birds.
g) Quality of the supply of drinking water: villagers gave a low score, 3.8, to the quality of the pressure, 4.3 to the chlorination system, and 5.2 to the intermittency of the supply.
h) Tourism: this activity develops with the greatest intensity during the festival of the Assumption of the Virgin. During those days about 9 families sell food and provide lodging for the visitors.
i) Annual income: the average annual family income is between $2,400 and $3,600
j) External investments: there are currently 5 projects in the village: reconstruction of the church, building a baby football field, the ESUSCON project, the

management of citruses, and the Internet. At the same time, there were 3 projects awaiting funding which were: a cellular antenna, improve Route A-85, and adjust the social centre.
k) Electrical consumption
l) Electrical devices: a register was made of the number and types of electrical devices. The most common devices are televisions, water boilers, decoders, and refrigerators.
m) Energy Demand Management
n) Access: the only access to the village is Route A-85 which is unusable during the strong summer rains.
o) Telecommunications: at present, the village is connected to the Internet at the G-101 school and a private telephone is installed in one of the residences. These are the only mediums of communication for the villagers.

IV. Discussion and Conclusion

The small number of 80 inhabitants in Huatacondo, facilitated the process of collective decision-making. This also made the delivery and explanation of information easier during the implementation of ESUSCON. In addition, it is worth noting that the integration of the Huatacondo villagers during the process permitted, together with the other activities described above, the application of different tools of participation which increased the possibility of the success of the project and the level of empowerment of the population in relation to ESUSCON. This contributed to the establishment of a dialogue and a permanent relationship between the community and the team in charge of the project that derived from the approval and acceptance of ESUSCON on the part of the majority of the Huatacondinos.

Amongst the activities carried out, the results obtained from the semi-structured surveys on characterization (done on more than 80% of the population) and the community perception of potential impacts (done on more than 60% of the population), helped form a significant view of the community facing these aspects. It is worth noting that the Huatacondinos collaborated actively during the entire process of implementation, contributing their knowledge about the village and ideas that were highly welcomed by the technical team. With respect to the workshops that were held, these allowed the community to make decisions on aspects such as the definition of the placement zones of the electric power generating units, the assignment of a name to the project, and the interface of the Energy Demand Management, among other things.

On the other hand, thanks to conversations with adults and young people, the Huatacondinos were able to avoid a series of negative impacts that the project might have generated, and they also brought awareness of other impacts that had not previously been considered, such as the replacement of jobs that require time to identify. They also suggested activities during the implementation phase that had not been contemplated in advance. Furthermore, they identified the benefits that would come with having electricity for 24 hours a day. According to the perception of the villagers, this would permit the realization of projects and ideas they had thought about, such as promoting tourism, developing crafts, and strengthening the "Center of Mothers". In this way, the project could improve the quality of life of the population in agreement with the definition of Gómez Orea, and raise the level of income in order to satisfy basic necessities[18]. This can be done once the activities requested by the community are in place and generate income.

It can be said that some points in favour of this project correspond to their capacity to be able adapt to different local realities and the generation of local and national benefits. This is because the distributed generation is not restricted to one single type of energy in particular, which permits development in the whole country. Furthermore, this type of project contributes to the diversification of the energy matrix, using natural resources of each particular area in a sustainable manner, and decentralizes energy that currently 72% of the population[19] depend on from energy sources which Chile imports.

Although the economic and environmental factors are important, there is no reason to set aside the social component because acceptance by the people is a fundamental factor in the success of projects of this nature[2]. Of great importance is the sense of belonging and right of the community to participate in, with respect to their geographical space, the types of economic activities established, the use of the region's natural resource areas and particularly the relationship they have with the territory. This needs to be realized before any type of intervention can take place where the community is involved in making decisions. It must include opinions that they have with respect to what will be developed.

The basic conditions for developing this type of project centre fundamentally on understanding the local realities in the area of intervention, without generating territorial, social, economic and environmental transformations that might convert the project into a negative element for the community. As a result, empowering and understanding the functioning of the community and the relationships existing within it becomes a fundamental factor by means of which the project can be sustained over time. The job done by and for the community through a committed way of thinking may raise the confidence amongst the people in organizations, and in this way local capacity is increased for future and collective actions too. It is important to note that social confidence in renewable energy projects is constructed using a bottom up method[20].

V. References

[1] R. Raven, R. Mourik and Y. Feenstra (2007). "Modulating societal acceptance in new energy projects".4th Conference on Sustainable Development of Energy Water and Environmental Systems, June 2007, Dubrovnic, Croatia.

[2] A. Krishnaswamy "Participatory Research: Strategies and Tools". Practitiones. Newsletter of the National Network of Forest Practitioners, vol. 22.

[3] R. Palma (2008). "La Iniciativa GeVi: El primer Generador Virtual para Chile". Dissertation, Dept. Electrical Eng., University of Chile.

[4] F. Sabatini y C. Sepúlveda (1964). "Conflictos ambientales: Entre la globalización y la sociedad civil". Santiago, Chile, CIPMA, 383p.

[5] G. Foster (1964). "Las Culturas Tradicionales y los cambios Técnicos". México, Fondo de Cultura Económica, 260p.

[6] R. Eyben, N. Kabeer and A. Cornwall (2008). Conceptualising empowerment and the implications for pro growth: A paper for the DAC Poverty Network. Septiembre de 2008. 37p.

[7] R. Brand and A. Karvonen. "The ecosystem of expertise: Complementary knowledges for sustainable development". Articule of Sustainability: Science Practice and Policy. [Online]. Available: http://ejournal.nbii.org

[8] R.B. Cialdini. "Crafting normative messages to protect the environment. Current Directions in Psychological Science" 12, pp.105-109.

[9] J. Riley (2001). "Multidisciplinary indicators of impact and change: key issues for identification and summary". Agriculture, Ecosystems & Environment, UK, 87, pp. 245-259.

[10] M. Reed, E. Fraser and A. Dougill. "An adaptive learning process for developing and applying sustainability indicators with local communities". Sustainability Research Institute, School of Earth and Environment, University of Leeds, United Kingdom. Available at www.sciencedirect.com

[11] Freebairn and C. King (2003). "Reflections on collectively working toward sustainability: indicators for indicators!" Australian Journal of Experimental Agriculture 43, pp. 223-238.

[12] S. Bell and S. Morse (2001). "Breaking through the glass ceiling: who really cares about sustainability indicators?" Local Environment: The international Journal of Justice and Sustainability, 6, pp. 291-309.

[13] N. Pretty (1995). "Participatory learning for sustainable agriculture". International Institute for Environment and Development, London, U.K, 23, pp. 1247-1263.

[14] T. Alberich (2007). "Investigación – Acción Participativa y Mapas Sociales". Benlloch (Castellón). Noviembre de 2007. 27p.

[15] M. Gutiérrez. "Mapas Sociales: Método y ejemplos prácticos". s.a. [Online]. http://www.fejidif.org/Herramientas/Otras/Gestion/DIAGNOSTICO%20ASOCIATIVO/mapassociales.rtf
[16] Asociación de Proyectos Comunitarios (A.P.C.) (2005). "Fortalecimiento de las organizaciones pertenecientes a la Asociación de Proyectos Comunitarios. A.P.C". Popayán, Colombia, 9p.
[17] Universidad Central de Chile. Apuntes "metodología de evaluación de paisaje". p. 9.
[18] D. Gómez (2002). "Evaluación de Impacto Ambiental" (2nd ed.) Barcelona, Aedos, p. 745.
[19] Comisión Nacional de Energía (CNE) (2006). Plan de Seguridad Energética (PSE). Santiago, Chile, 7p.
[20] G. Walker et al. (2009). Trust and community: Exploring the meanings, context and dynamics of community renewable energy. Energy Policy, doi: 10.1016/j.enpol.2009.05.055

Part B

Wind Power Scenario for Brazil

N.J. de Castro, G.A. Dantas, A.L.S. Leite[1]

Abstract

Recently, there has been much debate about the introduction of wind power into the electricity system. Wind power has the advantage of being a renewable energy source with low variable costs, but with very high fixed costs and investment costs. Thus, the purpose of this paper is to examine wind power costs in Brazil and the prospects for this energy source in Brazil's electric power mix, particularly after the wind energy auction held in late 2009. Wind power complements hydropower and can therefore be an important tool in guaranteeing system reliability. We also conclude that wind power is still one of the most expensive sources of electricity in Brazil, necessitating a specific policy for the development of new wind plants. One example of such a policy is the wind energy auction held in December 2009.

I. Introduction

With hydropower accounting for 90% of its electricity mix, Brazil's position is unique and privileged compared to the global electricity mix. This profile assures it a supply of competitive and clean renewable energy. However, it is only because of large reservoirs regulating electric power supply over the course of the year that Brazil is able to meet its electricity demand using hydrogeneration. Now, geographical, legal and environmental concerns prevent further construction of large reservoirs. As a result, Brazil's electricity mix will increasingly require the introduction of other energy sources to operate at the base in order to complement hydrogeneration during the dry season.

Accordingly, complementation of its hydro capacity should place higher priority on sources which are well suited to operating at the base, that is, sources with low variable costs. These options include low variable cost thermal plants, including coal-fired plants, thermal plants under inflexible contracts and, in particular, biomass and wind power plants. The latter two options offer the advantage of being natural complements to both hydrogeneration and renewable energy

---

[1] N.J. de Castro, *UFRJ*; G.A. Dantas, *UFRJ* and A.L.S. Leite, *UFFS*.

sources, and therefore compatible with the world goal of reducing greenhouse gas emissions[1, 2].

As a result, there is a clear need to promote investments in renewable sources of electricity, which for many years have been given secondary status in Brazil due to the abundance of hydro resources. However, when analysed on a short-term basis, such alternative sources tend to entail higher costs than conventional electricity generation sources.

Therefore, the purpose of this paper is to examine wind power costs in Brazil and the prospects for this energy source in Brazil's electric power mix, particularly after the wind energy auction held in late 2009.

This paper is divided into three more sections. In the next section, the analytical focus is on the growing need to complement Brazil's hydroelectric system and the importance of wind power as a complement. The third section discusses the costs, the results of the wind energy auction and the outlook for wind energy in Brazil. The forth section discusses the main conclusions.

II. Brazil's Energy Mix in Transition

In terms of its unique supply of primarily hydro-based electricity, Brazil can only be compared to a small number of other countries, including Norway, Canada and Venezuela. Table I shows that, in recent years, hydro resources have accounted for an average of 90% of electric power generation in Brazil.

Table I: Hydroelectricity in Proportion to Total Generation (%)

Year	Percentage
2000	94.11
2001	89.65
2002	90.97
2003	92.14
2004	88.63
2005	92.45
2006	91.81
2007	92.78
2008	88.61

Source: Brazil's national grid operator (www.ons.org.br)

The hydro base of Brazil's generating system ensures a supply of competitively priced power, while the auction results for plants on the Madeira River demonstrate that this is a continuing trend and a "clean" one, given the low carbon emission levels of hydro generation. This characteristic significantly contributes to the low carbon intensity of Brazil's energy mix, which is only 1.57 tons of CO_2 per TEP, while the intensity of the world energy mix is 2.36.

However, this predominantly hydroelectric generation exists along with highly irregular, strongly seasonal rainfall patterns. The specific characteristic that has enabled Brazil's electricity system to expand and meet demand all year round using a hydro base has been the construction of large reservoirs. This has enabled the electric power supply to be regulated throughout the year by stocking energy in the form of water during the wet season (stored energy) to be converted into electricity during the dry season.

However, although Brazil exploits only about 30% of its hydro potential, the model based on hydroelectric plants is unsustainable for two reasons. The first constraint is geographical: the hydroelectric potential of upland areas has already been harnessed and the remaining potential is located in flat regions of the country unsuited to the construction of regulation reservoirs. The second is legal: the more rigid environmental provisions of the 1988 constitution place restrictions on the construction of new large reservoirs and even hinder expansion of hydro-generation capacity based on run-of-the-river plants.

Transition to a reconfigured electricity mix in Brazil will require energy sources operating at the system base to complement hydro capacity during the dry season. The issue that arises is how to define what energy sources should be considered high priorities to complement hydro capacity for basic planning purposes. An analysis based on technical and economic variables would recommend opting for plants suited to operating at the base, that is, plants with limited technical flexibility and low variable costs, even though this may entail high investment costs, which would demand and warrant more hours of operation to amortize the investment. In this regard, in Brazil's current electricity system, priority is to be given to inflexible thermal plants (in particular to bioelectricity generated in sugar power plants), thermal plants with low variable costs and wind power.

The complementarity of wind power is illustrated in Fig. 1, which shows the relationship between rainfall and wind regimes in the Northeast, specifically in Ceará state. Note that the two regimes are highly complementary, making wind power a strong option for electricity generation during dry periods.

Sugar-powered bioelectricity and wind power should be priority energy sources for expanding Brazil's generating system, because not only are they natural complements to hydro generation; they also offer the added benefit of being renewable sources. Bioelectricity is complementary because the sugarcane harvest takes place between the months of May and November, which is the dry season in the mid-western region, where 70% of Brazil's reservoir capacity is located. Wind power, meanwhile, is complementary in that winds are more intense and consistent, precisely during the dry season of the year, especially in the northeast region, which offers Brazil's greatest wind power potential.

Fig. 1: Hydro and wind power complementarity

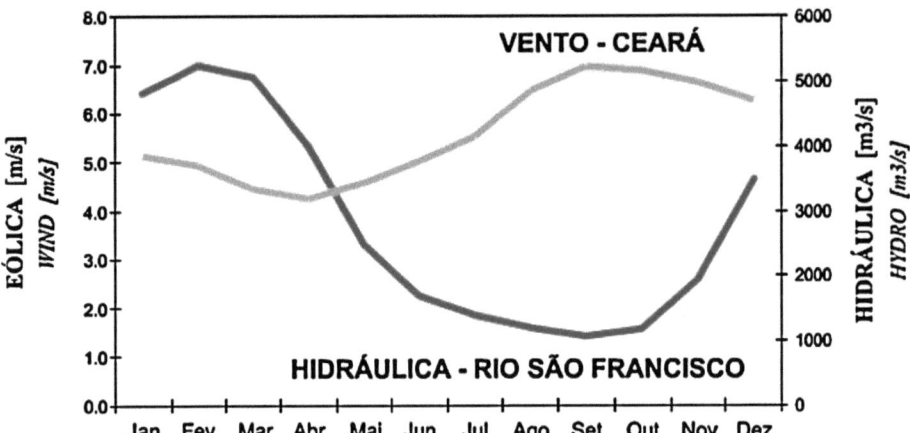

Source: H. Chipp. Procedimentos Operativos para Assegurar o Suprimento Energético do SIN[3]

III. Competitiveness of Wind Power and Policy Making in Brazil

Given the predominance of hydroelectricity in Brazil's electricity mix and the potential diversity of that mix, the main reason for promoting alternative, renewable electric power sources is different from the situation in other countries. In Brazil, the higher cost of wind power is an obstacle to its use, so it becomes fundamentally important that the government develop public policies to promote wind power.

The first part of this section discusses wind power costs in Brazil as compared to other countries. It then analyses the results of the wind power auction and the instruments the government used to make these results possible. Lastly, it discusses the need to continue and refine policies to promote wind power.

A. Costs and Competitiveness of Wind Power in Brazil

The gains of scale and economies of learning derived from the expansion of wind power generation throughout the world in recent years have enabled a significant reduction in costs. At present, the investment cost of wind technology stands at around US$1,900 per kW installed in the US; and even in countries where costs are higher, such as Germany, it is around US$2,000. However, investment in wind power in Brazil is expensive, with market estimates indicating values of US$3,000 per kW installed, according to[4].

Until mid 2009, these higher costs of wind power could be ascribed to Brazil's deficient infrastructure in certain aspects critical to the development of wind power ventures and the level of supply from the wind generator industry for Brazilian projects: only two manufacturers have facilities in Brazil and there are restrictions on the importing of equipment.

Brazil's infrastructure bottlenecks and their effects, are well known. These constraints place a burden on the manufacture of many products and make Brazil's economy less competitive to the point that this burden is known as the "Brazil cost". In the case of wind power ventures, not only do these bottlenecks increase investment costs, at the limit they can even make it unfeasible to build some projects. The precarious state of highways in the northeast region is the best example of the structural deficiencies in Brazil which have an adverse impact on wind power development. It is there that the greatest potential for wind power generation can be found, as shown in Table 2. However, the logistics of transporting one-hundred-metre-tall towers in such adverse conditions is a major cause for concern.

Table 2: Wind Power Potential in Brazil

Region	Potential (GW)	Potential (%)
North	12.8	8.9
Northeast	75.0	52.3
Mid-west	3.1	2.1
Southeast	29.7	20.8
South	22.8	15.9
Total	143.40	100.0

Source: R. Costa, B. Casotti, R. Azevedo. Um Panorama da Indústria de bens de Capital Relacionados à Energia Eólica[5]

As regards the supply of wind turbines, the industry is oligopolised worldwide, with the four largest firms (Vestas, GE Wind, Gamesa and Enercon) accounting for more than 70% of the market. Accordingly, the expansion of competition in this industry is essential in order to reduce the prices of the machinery. In an industry characterised by constant technological innovation, both in the production process and the types of turbines, it is of the utmost importance that prospective entrants have access to credit to ensure that the process of innovation is not restricted to established firms. Other important means of promoting competition in this industry would include investing in research at university centres, in partnership with new investors, and attracting foreign firms with an interest in operating in Brazil, or at least guaranteeing these firms an open market.

Until late 2009, the structure of Brazil's wind turbine industry was highly concentrated, with only two producers (Wobben and Impsa). The prohibition on importing wind turbines with less than 1.5 MW capacity and the 14% import tax place restrictions on importing countries, thus guaranteeing greater market power to manufacturers established in Brazil. In addition to this import restriction policy, there is a structural factor in the Brazilian economy which provides an enormous competitive advantage to manufacturers established on Brazilian soil: the National Economic and Social Development Bank (BNDES), the main source of long-term financing in the Brazilian economy, requires that capital goods be purchased from national producers. Therefore, it is essential to promote competition in wind turbine supply in order to reduce the cost of investment in wind farm projects, and this type of competition is dependent on attracting new foreign firms to the Brazilian market.

It should be mentioned that, although reserving the market to Brazilian national manufacturers is an unacceptable practice because it allows these companies to exert market power, these firms' competitiveness must be assured as a necessary condition for the wind generator industry to develop, and for wind power to expand, in Brazil. The strategic nature of "energy goods", especially a promising renewable source such as wind energy, warrants measures to ensure that investors who opt to set up business in Brazil are competitive, and ultimately sends a positive signal to new investors to come to Brazil. However, the issue to be discussed is: on what basis should this guaranteed competitiveness be offered?[2]

The option of taxing equipment imports is reasonable as long as the resulting tax revenues are used in favour of developing a national wind turbine industry, particularly in research and development to permit technological innovation.

Another alternative compatible with the ongoing endeavour to achieve tariff moderation is to opt for policies based on relaxing tax requirements for local producers in order to guarantee competition on a least-price basis. The tax ex-

---

2 Competitiveness can be defined as a firm's ability to formulate and implement strategies for competition which afford it a lasting, sustainable market position[6].

emption to be discussed below in this paper, together with signals that Brazil intends to contract wind power continuously in years to come, is one of the most important policies designed to attract new manufacturers to Brazil.

However, the necessary structural change will come from developing the Brazilian wind power equipment industry. For that development to take place, there must be indications of a systematic policy of wind power contracts. That relationship was made evident by the decision by GE and Siemens to set up manufacturing plants in Brazil following the wind power auction.

Besides the issues connected with infrastructure deficiencies and the short supply of wind turbines, the cost of investing in wind farms in Brazil is heavily burdened by excessive taxes and the high cost of capital. In this regard, reducing the cost of such ventures will depend on tax relaxation policies and lines of financing under favourable conditions. It was the use of these types of instruments that made the wind power auction a success.

On the other hand, certain characteristics of Brazil's electricity system make for lower costs in introducing wind power when compared to other countries. As this is an intermittent source of energy, introduction of major wind power capacity into the generating system requires that the system offer considerable surplus capacity. As Brazil's generating system consists predominantly of hydroelectric plants, installed capacity is already significantly greater than peak demand. Meanwhile, Brazil's wind power generating potential tends to be located far way from load centres, making investment in a transmission network a necessity. In this regard, given the scope of Brazil's existing transmission system, it is expected that such costs will tend to be lower in Brazil.

It must be stressed that, although the investment cost of a wind farm in Brazil may be higher than the costs of equivalent ventures in countries such as the United States, China or Germany, the intensity and regularity of winds in Brazil mean that the break-even tariff required to make such enterprises feasible in Brazil is not much different than the subsidised tariffs applied in other countries.

The foregoing explains the higher cost of wind power in Brazil, as compared to other countries and the main reasons for that cost. However, in strictly economic terms, the crucial issue for evaluating the potential for introducing wind power into Brazil's energy mix is the comparison between the costs of wind power and other energy sources in view of the present methodology for new energy contracts.

To begin with, this entails comparing the costs of wind power to the costs of other energy sources, while taking investment costs and variable costs into consideration and ignoring the energy contract methodology for the moment. Table 3 shows the costs of different generation sources.

The data in Table 3 indicates that wind power does indeed call for a higher tariff than most of the other sources, because of its high investment costs, associated with a lower capacity factor than the other sources. In that regard, in a reference scenario strictly based on economic variables, it would be correct to opt for the other sources – hydro, coal, natural gas and biomass – rather than wind generation. However, that economic analysis shows that wind power is extremely competitive when compared to fuel oil plants and that it is therefore unwarranted that these fossil fuel plants should be contracted instead of wind ventures, which is exactly what was happening in recent auctions. Therefore, wind power companies were uncompetitive in recent energy contract auctions not as a result of their costs, but rather because of the methodology used to appraise ventures in new energy auctions.

Table 3: Comparative Analysis of Generation Sources in Brazil

Sources	Investment (US$/kW installed)	Capacity factor (%)	Break-even tariff (US$/MWh)
Hydroelectric ("structural")	1,600	60	63
Hydroelectric	2,400	55	89
Coal	2,400	90	93
Natural gas	1,300	90	94
Bioeletricity	1,900	60	104
Small HEPs	3,600	55	112
Wind	3,600	35	119
Fuel oil	1,600	90	238

Source: Estimates prepared by GESEL/IE/UFRJ

The methodology currently used at new energy contract auctions is making fossil fuel thermal plants extremely competitive and, as a result, these types of plants are winning contracts. Briefly and without going into the discussion of the methodological problems of the cost-benefit index (ICB)[7], these plants are contracted on an availability basis and for dispatch only during a small number of hours per year.[7] Therefore, plants with low levels of flexibility are not competitive at these auctions. For a hydroelectric system where these thermoelectric plants are dispatched only a small number of hours per year, that is, they serve as system backup, contracting these thermal plants on an availability basis is good business for the Brazilian electricity system.

However, in a system that will increasingly need plants operating at the base in order to complement hydro generation during the dry season, thermal plants are not the best option because, when dispatched for a considerably greater number of hours than foreseen by the ICB calculations, they constitute a serious threat to efforts to maintain moderate tariffs in Brazil's electricity system because of their high variable cost. CASTRO/Dantas/Woodward/Leite[8] shows that Brazil runs a substantial risk of tariff hikes from the increased contracting of high variable-cost thermoelectric facilities.

Table 3 also shows the extent to which contracting high variable-cost thermal plants is going to burden the system, as such plants are dispatched more frequently. In the short term, because of security procedures, these plants operate for more hours than foreseen in the ICB calculation; over the mid-term, however, it is the system configuration itself which will require prolonged dispatch from such plants. Note that 70% of Brazil's flexible thermal capacity incurs a variable cost of more than R$200 per MWh, a value which makes a series of wind power ventures feasible.

Even with development of the wind turbine industry, plus tax relief and tax incentive policies to make wind power feasible at a tariff of US$100 per MWh, wind power ventures would find it very hard to compete with fuel-oil thermoelectric facilities, as can be seen from the difficulties that bioelectricity and inflexible natural gas thermal plants face in new energy auctions.

In this regard, it must be stressed that considerable challenges have yet to be met in order for Brazil to gain substantial wind power capacity. Some of the elements that pose challenges for the introduction of wind power plants in Brazil are listed below.

> Short history of wind measurement: the historical series for Brazil's wind regime is relatively short, spanning only 30 years. It was not until the 1970s that wind measurement began in Brazil. As the historical series is relatively short, it is considered unreliable.
> The Brazilian Wind Atlas (*Atlas Eólico Brasileiro*) needs revising with measurements from 100-meter towers: the measurements listed in the atlas were taken at 50 m. However, modern wind turbines are 100 m tall, making a detailed study of winds at that altitude necessary.
> Brazilian entrepreneurs' recent experiences in this field: here the learning curve promises lower costs and, consequently, more competitive energy from wind sources

B. Public Policy and Prospects for Wind Power in Brazil

The costs of wind power in Brazil and the methodology used for contracting new energy make contracting wind power through conventional mechanisms unfeasible. Therefore, specific policies should be developed to promote wind power. Such instruments act on the demand side by creating specific contracting mechanisms, as well as on the supply side by reducing the costs of such ventures.

The wind energy auction held in December 2009 was a typical example of a demand-side instrument to promote a specific source. However, instruments of this type cannot ignore variable costs, meaning that they must be accompanied by supply-side instruments which act to reduce costs. It was a well prepared tax relief policy, together with tax incentives and a discount on the "cable tariff" specified for renewable energy ventures with a capacity of less than 30 MW, which – added to the prospect of cost reductions from the entry of new firms producing wind power equipment – allowed the auction to achieve the result of 753 MWmed contracted, which is equivalent to adding 1805.7 MW to the system at the considerable price of US$ 98.27.

That result should not be interpreted as a sporadic event, but rather as the start of a systematic policy of contracting wind energy. To that end, the supply-side incentives should be maintained so as to allow wind power to be introduced into Brazil's electricity mix, taking the prices quoted at the wind energy auction as the price parameter, which were quite close to the price level for conventional sources. It is important, however, to maintain the option of specific contracting instruments because, as a result of the methodological problems at the general auctions, as indicated above, there is the risk of wind power not being contracted, even when taken jointly with sugar- powered bioelectricity, which should be a priority source in expanding Brazil's electricity system.

To underscore the importance of public policies for promoting wind power and renewable sources in general, a brief discussion of the instruments commonly used for this purpose, and their applicability for Brazil, is outlined below.

Specific Contracting Mechanisms

The most direct and efficient way to promote renewable alternative sources is to implement specific programmes to contract energy from such sources, for example, how the feed-in method works. These methods are essential for promoting renewable alternative energy sources because they guarantee that a market is created for such sources while, in the absence of demand, supply-side policies alone may not prove effective in ensuring that power generated from such sources is purchased.

P. Komor[10] argues that the feed-in method is at odds with the policy of bringing competition to the electricity sector. This means, in a sense, that the state is resuming its central role in the electricity sector^3.

Which of the two methods should be adopted will depend on the particular conditions and institutional framework in each country. However, the idea that auctions

---

3 This issue, in the European context, is addressed in greater detail in [11].

are an instrument which involves less government intervention and yields more efficient results should be demystified. The extent of intervention in the two instruments is similar: for feed-in, the government sets the price to be paid for energy; at auction, the government determines the amount to be contracted.

International experience indicates that programmes based on feed-in are more effective than policies based on systems of auctions, which often attract a number of speculative bidders who do not actually implement projects. Rather, these bidders often participate with a goal of selling the project on or bidding for the project at an impracticable price in the expectation that generating costs will decrease. Dutra[11] reports that the auction system was used in the United Kingdom in the 1990s in line with economic liberalization prescriptions, but that the number of ventures contracted which actually went into operation declined at each auction, culminating in an implementation rate of only 33% at the latest auction. She attributes the fact that many projects were not designed with excessive emphasis on competition and a focus on the centralized nature of electricity sector planning in the UK, in addition to the opportunistic conduct of some agents – as described above – who quote prices at less than cost in the expectation that future technological development will make their projects workable.

An analogy can be drawn between the UK case and Brazil's Alternative Source Incentive Programme (*Programa de Incentivo a Fontes Alternativas*, PROINFA), implemented to contract 3,300 MW divided between biomass, wind power and small hydroelectric plants. The PROINFA was based on a hybrid auction and feed-in methodology, because the government set both the price and the amount of energy to be contracted in advance. However, implementation of the programme was hindered by a series of constraints, including an inadequate supply of wind turbines and difficulties in financing for small investors, added to the speculative behaviour of some agents and the difficulties involved in connecting some projects to the grid. As a result, the programme – which should have added approximately 1,400 MW of wind power generation by the end of 2006 – suffered successive delays and by September 2009, the total wind generation capacity installed was 547 MW, with forecasts of 1427 MW by the end of 2010.

The use of such instruments should be restricted to early development until scale of production, learning economies and technological innovations reduce the cost of these alternative technologies. In this respect, priority should be given to instruments that encourage technological development. A. Leite, N. Castro[12] notes that, by assuring investors of return, the feed-in system encourages technological innovation because it enables lower-cost producers to earn differential revenue. On the contrary, the auction system, by matching prices and producers' marginal costs, does not encourage technological innovation.

On the other hand, feed-in must not be allowed to have abusive effects on the tariffs charged to final consumers. Rather, mechanisms have to be set up to prevent the entrepreneur from appropriating all the revenue derived from preferred conditions of project siting, technological innovations and market development. It is therefore necessary to adopt mechanisms to make the feed-in system more flexible, as occurs in Germany, which introduced a decreasing feed-in system differentiated according to conditions based on where the plant is located.

Therefore, it is important to develop and introduce an instrument that actually values wind power appropriately, considering all the positive externalities – including those in the environmental dimension – that this source affords Brazil's electricity sector. At the same time, the structural and institutional conditions must be present to guarantee that the projects contracted are actually implemented.

Note that creating a green certificates market does not appear to be the best option for Brazil, because Brazil already has an essentially renewable electric power mix. Thus, there would not be much substance to the requirement that consumer portfolios include a percentage from renewable sources.

Therefore, in the Brazilian case, when there is an institutional framework based on auctions, these are the appropriate instruments for contracting wind energy. However, as mentioned earlier, these must be specific auctions, because the present state of development of wind power and the energy auction rules make it quite unlikely that wind ventures will win generic auctions, even given the supply-side incentives.

Special Lines of Financing

Larger and more conservative projects, especially when implemented by major corporations, manage to secure funding under advantageous conditions. Conversely, wind power projects, which inevitably are not large scale and are based on technology still under development, do not usually obtain funding from the banking system under ideal conditions.

In Brazil, the National Economic and Social Development Bank (BNDES) is the only source of long-term capital for project development. In one of its lines of funding, the BNDES should offer greater incentives to promote investment in efficient, innovative and sustainable technologies, which would include wind power[13]. However, the conditions of financing for alternative renewable sources are quite similar to those granted for projects based on conventional electric power generation technologies, as can be seen in Table 4.

Table 4: BNDES Financing Conditions for Electricity Generation Projects

Lines of financing	Cost of capital (% per year)	Maximum BNDES participation (%)	Amortization period (years)
"Structural" HEPs (installed capacity > 1,000 MW)	0.9 + TJLP	80	20
HEPs	0.9 + TJLP	80	16
Natural gas-fired, bio-electricity, wind plants and small HEPs	0.9 + TJLP	80	14
Oil and coal-fired TEPs	1.8 + 50% TJLP + 50% TJ-462	60	14

Source: http://www.bndes.gov.br

As shown in Table 4, there are no specific, subsidized lines of financing for alternative sources, and it is highly questionable that conditions should be the same for such sources as for hydroelectric and natural gas-fired thermal plants, in view of the larger scale and mature technology of these sources.

That subsidized lines of financing should be formatted for wind power is warranted by the strategic nature of wind energy. The granting of lines of financing under special conditions is an important instrument for promoting alternative, renewable energy and is a means of promoting the desired energy policy. In that regard, there should be a dialogue between the BNDES and the bodies responsible for electric sector planning, so as to define what sources should be prioritized and for financing to be offered at more favourable conditions.

IV. Concluding Remarks

We conclude that wind power plants are important to complement hydroelectric plants in Brazil's electricity sector. Wind power complements hydro power, and can thus be an important tool to assure system reliability.

We also conclude that wind power is still one of the most expensive sources of electricity in Brazil, necessitating a specific policy for the development of wind plants. One example of such a policy is the wind energy auction held in December 2009.

Lastly, we find that in order for Brazil to use its wind power potential fully, it should stimulate a domestic market for wind turbines. Some suggestions have been given here for effectively planning and exploiting RE resources and using the related technologies. These suggestions are useful not only for Brazil, but are equally important for other countries to significantly boost the RE contribution of their total energy supply.

References

[1] N. Boccard (2009). "Economic properties of wind power". Energy Policy, 10, p.1016.
[2] U. Büsgen (2009). "The expansion of electricity generation from renewable energies in Germany: A review based on the Renewable Energy Sources Act Progress Report 2007 and the new German feed-in legislation". Energy Policy 37, pp. 2536-2545.
[3] H. Chipp (2008). Procedimentos Operativos para Assegurar o Suprimento Energético do SIN. Presentation at GESEL-IE-UFRJ, Rio de Janeiro, 9 July 2008.
[4] M. Torres (2009). Desafios para o desenvolvimento do setor de Energias Renováveis no Brasil. Florianópolis: UNISUL (PowerPoint presentation), 10 June 2009.
[5] R. Costa, B. Casotti and R. Azevedo (2009). Um Panorama da Indústria de bens de Capital Relacionados à Energia Eólica. BNDES. Rio de Janeiro.
[6] O. Amarante, W. Brower and J. Zack (2001). A. Sá Atlas do Potencial Eólico Brasileiro. CRESESB/ELETRO"BRAS/CEPEL/MME. Brasília: MME.
[7] L. Coutinho and J. Ferraz (1994). Estudo da competitividade da indústria brasileira. Campinas: Papirus, 512 p.
[8] N. Castro, G. Dantas, J. Woodward and A. Leite (2010). "Perspectivas para a Energia Eólica no Brasil". Texto de Discussão do Setor Elétrico, no. 16. GESEL/IE/UFRJ.
[9] N. Castro and D. Bueno (2007). Os Leilões de Energia Nova: Vetores de Crise ou de Ajuste Entre Oferta e Demanda? Rio de Janeiro: IE-UFRJ, 18 June 2007.
[10] P. Komor (2004). Renewable energy policy. New York: Diebold Institute for Public Policy Studies.
[11] R. Dutra (2007). "Propostas de políticas específicas para energia eólica no Brasil após a primeira fase do PROINFA". PhD dissertation. COPPE, Universidade Federal do Rio de Janeiro.
[12] N. Castro, R. Brandão and G. Dantas (2009). Alternativas de Complementação do Parque Hídrico. Mimeo. GESEL/IE/UFRJ. Rio de Janeiro.
[13] A. Leite and N. Castro (2009). "Política para o setor elétrico da União Europeia". Economica, vol. 11, no. 2, pp. 111-132.

Energy Recovery from Biodegradable Waste in the Grain Processing Industry

J.K. Staniškis, I. Kliopova, V. Petraškienė[1]

Abstract

The management of biodegradable (BD) waste is one of the most important environmental problems in the grain processing industry. BD waste enters the production process with unprocessed raw materials; it is generated in every stage of grain processing until raw material becomes commodity. BD waste from grain processing is considered to be non-hazardous waste and is classified in category 02 03 of the European waste catalogue.

This paper is focused on analysis of the possibility of using BD waste for the production of solid recovery fuel. Cleaner production audits have suggested two significant aspects: (1) the generation of biodegradable waste in grain processing can be up to 18.3 kg/t of product; (2) energy consumption can be up to 108.3 kWh/t of product. Theoretical and experimental investigations in a pilot grain processing company have proved that the use of biodegradable waste for the production of alternative fuel in addition to energy would reduce the amount of BD waste by up to 4.48 kg/t of product.

The mass and energy balances of the processes mentioned above, plus environmental and economical benefits, are presented and discussed in the paper.

I. Introduction

Grain processors have been obliged to seek alternative methods to improve environmental performance as a result of rising energy prices, waste management costs, fees for natural resources, fines for pollution and the 'polluter pays' principle outlined in the Environmental Law of the Republic of Lithuania.

Up to now, some grain-processing (GP) companies were able to deliver solid biodegradable (BD) waste to waste management companies along with municipal waste; others composted BD waste or simply collected it on their sites.

---

1 Lithuania.

The generation of BD waste in large volumes is a feature of this industry. The BD waste problem can be solved by optimising GP technological processes through the application of new methods for making use of waste.

The main technological processes of GP companies that produce BD waste are the following:

- grain receiving and primary processing: grain cleaning and storage in elevators, drying and dried grain storage in bins;
- flour production: wheat and rye preparation, grain conditioning, milling, quality estimation;
- grits production: grain preparation (from buckwheat, barley, wheat, oats, pulses), processing (husking), cutting, improving, drying;
- mixed fodder production: grain preparation, milling, flour sieving, batching to the prescribed mixture, mixing to the mixed fodders, formation (for example, granulation, extruding);
- packing and loading of manufactured production.

Other GP processes:

- heat energy production (mostly in steam boilers fuelled by natural gas);
- treatment of air emissions in cyclones;
- BD waste management (collection, temporal storage, delivering to waste management companies or composting on company sites).

Table I: Material and bd waste-flows in a sample gp company (primary evaluation)[1]

Input and output materials of the grain-processing processes	2008 m. (t/year)
Input materials:	**109770.09**
Grain	62000
Additional raw materials	45769.89
Protein raw materials	2000.2
Output materials:	**108575.00**
Manufactured products:	107796
Highest and first-quality flour	12000
Mixed fodder	87000
Protein vitamins and supplements	2000
Farina	2500
Bran (by-products with about 70% of nutrients)	4256
Other products	40
Waste volume indicated in yearly statistical waste report for 2008	779
Difference between inputs and outputs (BD waste)	**1195.09**

The qualitative and quantitative character of waste surely depends on the quality of supplied raw materials, technology used, the state of technological equipment, the type and quality of manufactured production, climatic conditions during grain growth, etc.

The key goal of the research was to identify the most significant environmental aspects of grain processing, and – through the application of the Cleaner Production (CP) assessment method – to identify the best possible solution from an economic and environmental point of view. For this purpose: (1) the material and energy balances of grain processing were developed; (2) the main environmental aspects were identified and BD waste was recognised as the most significant form of waste in a sample GP company; (3) technical, economic and environmental benefits were evaluated using the Cleaner Production assessment method and energy recovery was accepted as the best solution for BD waste management; and (4) technical realisation of the selected solution was proposed.

II. Methodology

The methodology of the evaluation of the significant environmental aspects of grain processing is represented in Fig. 1.

As a first step, a questionnaire was developed and used to conduct the survey. This questionnaire was sent to all member companies of the Lithuanian Grain Processors' Association. Completed questionnaires were received from 24 companies, which represented a response rate of 75%. The main objective of the survey was the identification of significant environmental aspects and possible solutions.

Following analysis of the survey results, an environmental audit was carried out in the sample company to analyse the main environmental aspects and their environmental impacts. During the audit, the Integrated Pollution Prevention and Control (IPPC) permit was analysed, input and output flows of the main technological processes were measured together with the company specialists, and material and energy balances were made. Environmental indicators were used to evaluate the company's environmental efficiency. The main principles of Systems theory were applied to define the problem and the objectives for process optimisation.

Fig. 1. Methodology for the evaluation of environmental aspects and impacts and possibilities for increasing environmental efficiency in grain-processing company

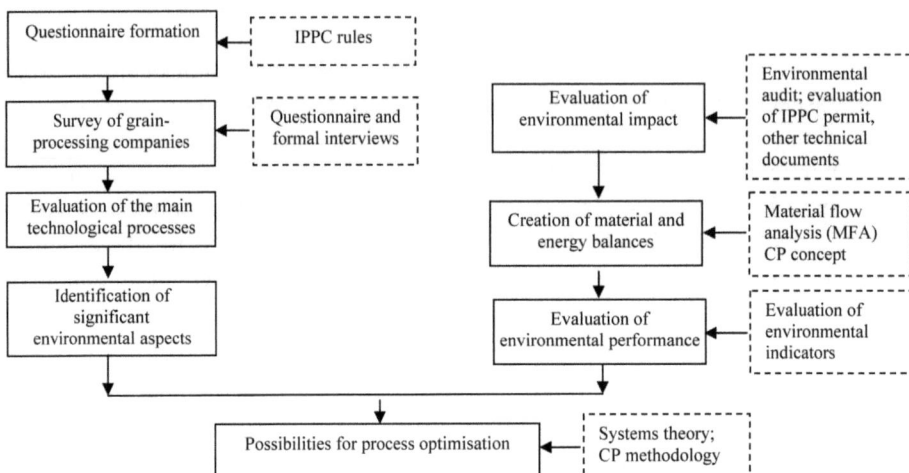

Cleaner Production methodology was used for the identification of environmental aspects and assessment of the performance improvement alternatives. Cleaner Production means a continuous application of an integrated preventive environmental strategy to processes, products and services to increase overall efficiency. This leads to improved environmental performance, cost savings, and the reduction of risks to humans and the environment:

- for production processes, CP includes conserving raw materials and energy, eliminating toxic raw materials, and reducing the quantity and toxicity of all emissions and waste before they leave the process,
- for products, CP focuses on reducing impacts along the entire life cycle of the product, from raw material extraction to the ultimate disposal of the product.

For services, using a preventive approach involves design issues, housekeeping improvement, and a better selection of material inputs (in the form of products)[2, 3].

III. Environmental Impact of Grain Processing

Input and output flows of typical grain processing equipment are presented in Fig. 2. Amount and toxicity depend on many factors. For instance: manufactured products, technological processes used, technical state of equipment, choice of control equipment or control system, environmental decisions applied, personnel competence, and other related factors.

During the survey, the respondents from 24 Lithuanian GP companies had to indicate the main environmental problems they face, identify the reasons for these problems, and list measures used to combat them. The respondents indicated that the main environmental problems in grain processing are the following: BD waste generation and management (100% of companies), non-efficient energy consumption (100%), waste water sludge (80%) and emissions (80%)[1].

The analysis results of the GP industry production balances and the statistical data show that in 2007 more than 64 thousand tons of BD was generated in flour and grits production alone. For further analysis in the sample company, the process input and output data were estimated using all the available information sources and measurement methods. An inventory was made of the material and energy flows entering and leaving the company with the associated costs. A process diagram was then drawn up, allowing the identification of all sources of waste and emissions. Material flow analysis (MFA) is a systematic assessment of the flows and stocks of material within the system defined in space and time. It connects the sources, pathways and intermediate and final sinks of material[4, 5].

The material and energy balance based on MFA analysis has quantitatively proved that the main environmental aspects of grain processing are BD waste generation and non-efficient energy consumption. Since those two main aspects are interrelated, solutions were sought for both of them together.

A detailed analysis of the solutions/options that were brainstormed revealed that solid recovered fuel production is the most promising way of managing biodegradable waste from grain production and increasing energy efficiency.

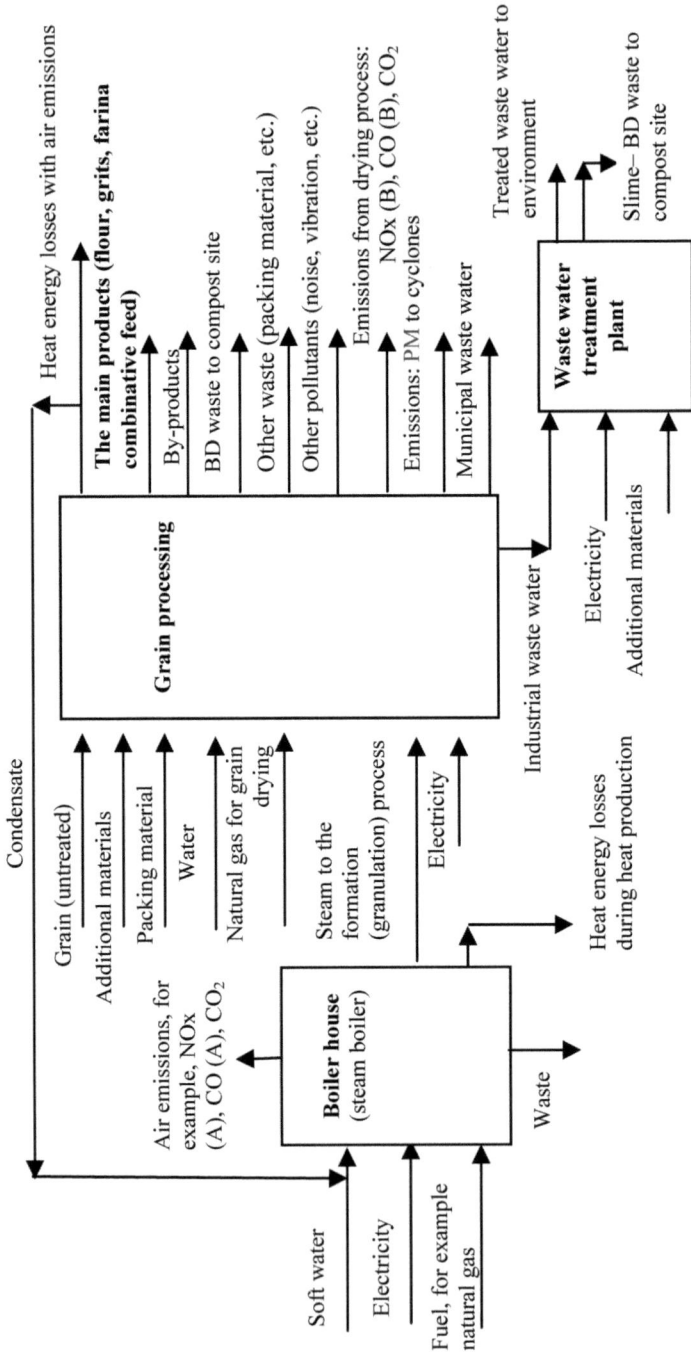

Fig. 2 Flowchart of typical grain-processing company (PM – particulate matter)

IV. Feasibility Analysis

A. Technical Assessment

The composition of solid BD waste in the sample GP company is as follows:

- grain impurities: split or sprouted grain, mixes with other grain, affected by vermin etc.;
- waste impurities: other seeds, mineral or organic additives, insect remains;
- grain peels;
- bran;
- solid particles from emissions treatment equipment (cyclones).

Processing of solid BD waste (separated inert material, milled, homogenised, pressed or extruded) is necessary before incineration. BD waste prepared in this way becomes solid recovered fuel[6].

The solid recovered fuel production experiment was carried out in a company producing wood pellets. The pellets were produced from solid BD waste from the sample GP company. A press with a capacity of 1.1 t/h was used for this purpose. Pellets (6.00 mm in diameter and 1.5-2.0 cm in length) were successfully produced from different BD waste materials without additional watering or drying. The moisture in the pellets produced did not exceed 15%.

To determine the physical and chemical characteristics of the pellets produced, a laboratory analysis was carried out. The following characteristics of the recovered fuel were determined: moisture (%), net calorific value as received (MJ/kg), net calorific value in dry matter (MJ/kg), ash content in dry matter, sulphur (S), nitrogen (N), carbon (C) contents in dry matter (%), heavy metal content in dry matter (%), iron (Fe), calcium (Ca), magnesium (Mg), natrium (Na), kalium (K), aluminium (Al) contents in dry matter (%). The classification system for solid recovered fuel, as presented in CEN/TC 343, is based on three main fuel characteristics[7]:

- net calorific value (as received – ar);
- chlorine (Cl) content (in dry matter – d);
- mercury (Hg) amount (as received – ar).

According to the categories presented above, the recovered fuel produced during the experiment was examined and classed. The results are presented in the Table 2. In the sample company, approximately 5.9 thousand t of BD waste and around 3.76 thousand t of unsold bran could be used as raw material for pellets production. A pellet press (up to 1.7 t/h of capacity; 125 kW of electric capacity; matrix with 8 -14 mm of diameter) is needed for this purpose. For heat energy production, an automated modern boiler (with 90-125 kW of heat capacity) for burning solid fuels in pellets should be installed. The efficiency of such a combustion plant would be about 85-89%. Measurable ash-content of recovered fuel (up to 8-10%) was observed during laboratory analysis. Therefore, centrifugal precipitation (cyclones) is required to remove solid particles from the flue gas.

Table II: Standardisation of solid recovered fuel produced, according to the classification system of solid recovered fuels (CEN/TC 343)

BD waste of GP company	Conformity to a certain class of recovered fuel according to fuel characteristics presented below			
	Net calorific value (ar), MJ/kg	chlorine (Cl) content (d), %	Mercury (Hg) content, mg/MJ (median)	Mercury (Hg) content, mg/MJ (in 80%)
Pellets from composite solid BD waste of GP	4th class (≥10)	1st class (≤0,2)	1st class (≤0,02)	1st class (≤0,04)
Pellets from bran	4th class (≥10)	1st class (≤0,2)	1st class (≤0,02)	1st class (≤0,04)

B. Environmental Assessment

To evaluate all input and output flows of the sample GP company's technological processes (incl. recovered fuel production, heat energy production, burning recovered fuel etc.), the material and energy balances were developed for the situation "after innovation". Target absolute values (units/year) and relative values (units/t of manufactured products) of the environmental indicators were determined from the balances and compared to the values existing "before innovation" (see Table III).

The implementation of an energy recovery project will enable the solution of the main GP company's environmental problems; i.e. solid BD waste will become raw material for recovered fuel production – an alternative heat energy source. Partial substitution of the recovered fuel produced for natural gas will allow CO_2 emissions from the company's stationary sources to be reduced by approximately 1 thousand t/year (see Table III).

The production of recovered fuel added to a new boiler house will mean the company's electricity consumption will increase by 447 MWh/year. At the same time, CO(A) emissions in exhaust gas will increase by 17.63 t/year, and solid particles by 66.08 t/year. Furthermore, new BD waste will be generated through losses during recovered fuel production and PM from cyclones.

Table III: Evaluation of environmental and economic benefits of recovered fuel production from BS waste in GP company

Input and output flows	Environmental indicators before innovation (existing situation)		Environmental indicators after innovation (planning)		Savings	
	Absolute value, units/year	Relative value	Absolute value, units/year	Relative value	Absolute value, units/year	EUR/year
Grain products (GP)	107 796 t, inc. 3 756 t of bran		104 040 t (without bran)		New product will produce – 5 386 t (fuel)	
Recovered fuel (RF)	-		5 386 t, inc. 1 428 for own purposes			
Total manufactured production (MP):	107 796 t, inc. 3 756 t of bran		109 426 t		Increase of MP volume by 1 630 t	
BD waste	1 954 t	18.13 kg/t of MP	466 t	4.26 kg/t of MP	1 488 t	27 441
Diesel fuel (for BD management)	0.57 t	0.01 kg/t of MP	-	-	0.57 t	3 425
Electricity (total)	4 418 MWh	40.98 kWh/t of MP	4 865 MWh	44.46 kWh/t of MP	-447 MWh	-49 195
Indirect environmental impact due to electricity consumption CO NOx CO_2	1.29 1.78 995	9.26 kg/t of MP	1.42 1.9 1 095	10.04 kg/t of MP	-0.13 -0.12 -100	-
Heat energy (total)	7 256 MWh	67.31 kWh/t of MP	7 256 MWh	67.31 kWh/t of MP	-	-
Natural gas (for heat energy production)	819 904 nm^3	7.61 nm^3/t of MP	289 517 nm^3	2.65 nm^3/t of MP	530 387 nm^3	230 416
Recovered fuel produced (for heat energy production)	-	-	1 428 t	13.05 kg/t of MP	-1 428 t	-
Air emissions during fuel burning: CO NOx PM: CO_2	2.02 t 2.75 t - 1 562 t	14.53 kg/t of MP	19.65 t 9.09 t 66.08 t 552 t	5.91 kg/t of MP	-17.63 t -6.34 t -66.08 t 1 010 t	-77 -1 249 -12 450 -
Employees (for BD waste management)	1	-	4	-	-3	-28 151
Additional spare parts for RF production (matrix)	-	-	2.7 units	-	-2.7 units	-15 640
PM emissions during RF production	-	-	0.17 t	0.03 kg/t of RF	-0.17 t	-33
Laboratory analysis of RF	-	-	12 units	-	-12 units	-3.128
Incomes due to RF purchasing	-	-	3 958 t	-	3 958	189 142
						340 501

Total environmental benefit (in absolute value):

- reduction of company's organic waste by 1 488 t/year (without bran);
- production of alternative energy (4 562 MWh/year);
- reduction of natural gas consumption by 530 thousand nm^3/year;
- direct and indirect reduction of air emissions by 819.7 t/year.

C. Economic Assessment

The energy recovery of BD waste in the sample GP company will enable a reduction in the cost of direct processes of up to 151.4 thousand EUR/year. Estimated income from recovered fuel sold: 190 thousand EUR/ year[8] (see Table III). Total project investment, incl. *design, press for pellets production, water-heating boiler, equipment transportation and installation, start–up adjustment*: 280 000 EUR. Payback period: 0.8 years.

V. Conclusions

1. It was determined that the generation of biodegradable waste in grain processing (up to 18.13 kg/t of manufactured production) and high energy consumption (up to 108.3 kWh/t of production) are currently the most significant environmental aspects of this industry.
2. The use of biodegradable waste for the production of alternative fuel would reduce the amount of biodegradable waste by up to 4.48 kg/t of grain processing products in the sample GP company. Consumption of natural gas would be reduced by 62.87% due to the combustion of part of the recovered fuel produced for heat energy production. Air emissions of combustion products would be reduced by 8.62 kg/t of the manufactured production.
3. In Lithuania, approximately 95 thousand t/year of solid recovered fuel can be produced from biodegradable waste from grain processing with a net calorific value comparable to the calorific value of dried sawdust (with up to 25-30% of moisture content). Up to 42% of this fuel can be combusted in the company's own boiler house to produce heat energy for heating, primary grain processing, grain milling, mixed fodder production and other processes involved in grain processing. After evaluation of the relevant chemical and physical characteristics, a further portion of the recovered fuel produced can be successfully sold as an alternative local energy source, reducing the payback period of investments by up to one year.

4. Partial substitution of the recovered fuel produced for natural gas in the grain processing industry will allow the production of up to 300 thousand MWh/year of alternative energy. The environmental impact of heat energy production on the atmosphere will therefore be reduced by approx. 67 thousand t/year (including CO_2 emissions).
5. The positive impact on the environment and the real economic benefits of the suggested method of biodegradable waste management should encourage grain processors to start developing new waste management practices and introducing alternative energy production.

VI. References

[1] I. Kliopova and V. Petraškienė (2009). "Evaluation of significant environmental aspects in Grain Processing". Environmental Research, Engineering and Management. Kaunas, Technologija 49, pp. 44-55.
[2] I. Kliopova and J.K. Staniškis (2003). "The evaluation of Cleaner Production performance in Lithuanian industries". Journal of Cleaner Production. Elsevier Science, vol. 11, pp. 619-628.
[3] I. Kliopova and J.K. Staniškis (2004). "Process Control in Cleaner production". Environmental Engineering and Management Journal. Technical University of Iasi, Ah.Asachi. 4, pp. 517-527.
[4] J.K. Staniškis, Ž. Stasiškienė and Chr. Jasch (2008). Cleaner Technologies: Environmental Management Accounting, Investment Appraisal and Financing .New York, Nova Science Publishers, Inc., p. 345.
[5] J.K. Staniškis and C. Jayaraman (2010). Cleaner Production and Energy Conservation for Sustainable Development. Delhi. Daya Publishing House, p. 288.
[6] S. Grubliauskas and N. Pedišius (2007). "Standardization of soil recovered fuel". Energetika Journal of Lithuanian Academy of Sciences 3, pp. 57-61.
[7] European Committee for Standardization (2006). Technical Committee 343 Solid Recovered Fuels (CEN TC 343). CEN/TR 15508:2006. Key properties on solid recovered fuels to be used for establishing a classification system, p. 78.
[8] G. Thek and I. Obernberger (2009). "Wood pellet production costs under Austrian framework conditions" in The 17th European Biomass Conference & Exhibition: proceedings. Hamburg, ETA-Renewable Energies, Italy, pp. 2129-2138.

A Study of Voltage Dips and Disturbances in Spanish Photovoltaic Power Plants

J. Guerrero-Pérez, F. Espín, J. Martínez,
A. Molina-García, E. Gómez Lázaro[1]

Abstract

Some European transmission network operators are proposing additional grid connection requirements for renewable power plants, particularly wind farms. These requirements will be extended to photovoltaic plants during the forthcoming years to include this renewable source as a significant power contribution avoiding undesirable interruptions from voltage dips or disturbances. Within this framework, the paper discusses a preliminary study of voltage dips and disturbances measured in Spanish photovoltaic power plants. This analysis gives additional information about the requirements demanded by the new generation of inverters and the minimum characteristics they should offer in a mid-term. Real measurements have been recorded in Spanish photovoltaic plants over several months, from both AC and DC sides. A full description of the results, as well as a detailed analysis of the disturbances in comparison with previous data collected in Spanish wind farms, are also included in the paper.

I. Introduction

Power supply from photovoltaic (PV) power plants has increased considerably over the past years, mainly due to the development of new solar panel technologies as well as different policies and incentives offered in several countries. In this case, PV plants will become a relevant influence on the operation of power systems in the near future. As the operation of power systems is supported by grid codes producing a set of requirements for all network users to comply to, some European transmission network operators have recently introduced special and additional grid connection requirements for renewable power plants, particularly wind farms requiring uninterrupted generation throughout power system dis-

---

[1] J. Guerrero-Pérez, *Gehrlicher Solar*, F. Espín, *Gehrlicher Solar*, J. Martínez, *Gehrlicher Solar*, A. Molina-García, *U.P. de Cartagena*, E. Gómez Lázaro, *U. de Castilla La Mancha*

turbances related to the network voltage. These requirements are likely to be extended to PV plants during the following years in order to include this renewable source as a significant power contribution and avoid undesirable interruptions of this generation under voltage dips or disturbances. Additionally, the large amount of small-scale generators being connected to the electricity distribution network at low voltage (LV) level could also affect voltage reliability problems; voltage quality supplied to customers is set at European Standard EN 50160[1].

Within this framework, the ride through requirements imposed on power plants is one of the key aspects of the grid codes, since premature tripping of those plants due to power system disturbances can put the stability of the system at risk, contributing to the amplification of the disturbance,[2-7]. Significant technical issues have to be considered by the inverter's manufacturers in a mid-term by adjusting their products to the grid code requirements. In this case, preliminary studies of voltage dips and disturbances in PV plants are thus desirable to offer a brief idea about the most common disturbances and transients currently suffered from this kind of power generation. This analysis offers additional information about the requirements demanded by the new generation of inverters. With this aim, real measurements have been recorded in Spanish PV plants over several months, taking voltage data from both AC and DC sides. A full description of the results, as well as a detailed analysis of the disturbances in comparison with previous data collected in Spanish wind farms, are also included in the paper.

The rest of the paper is structured as follows: Section II describes the voltage dip characterisation processes previously proposed in the relevant literature. Section III discusses the real data measurements recorded in different Spanish PV plants, describing the data acquisition system as well as different examples of the measured voltage dips. Section IV describes examples of voltage dips measured in Spanish wind farms, considering both duration and severity of the disturbances, and compares these with the voltage dips collected in PV plants. Finally, Section V gives the most relevant conclusions of the paper.

II. Voltage Dip Characterisation

Voltage-dip characterisation concerns the quantification of voltage-dip events through a limited number of parameters[8]. Most methods of voltage-dip characterisation use two parameters to quantify the severity of the disturbance: magnitude and duration[9-11]. In M. H. J. Bollen, *"Algorithms for characterizing measured three-phase unbalanced voltage dips,"*[9] two methods to obtain three-phase voltage-dip characterisation – 'ABC classification' and 'symmetrical components classification'– are exposed and compared. It is concluded that ABC classification, due to its simplicity, is more widely used than the symmetrical components, also being more intuitive and giving a good approximation about the evolution of the dips along the different voltage levels of the network. The ABC classification should not be viewed as a different classification. It must be considered as a special case of the symmetrical components classification. For our purposes, the previous characterisation processes appear too complex, since it is necessary to characterise the voltage dip, starting with its evolution through time and the end of the voltage dip at the point of measurement. For this reason, the voltage-dip representation is then based on the information of the phase voltage and current magnitudes along the disturbances. As additional information, Fig. 1 shows the requirements for voltage-operating ranges in different countries[12], expressed as RMS voltage in percent. Spain, Denmark, Germany, Ireland, Sweden and Scotland are summarised in Fig. 1, where their requirements related with both under-voltage and over-voltage protections as well as islanding situations are discussed.

Fig. 1: Requirements for voltage operating range

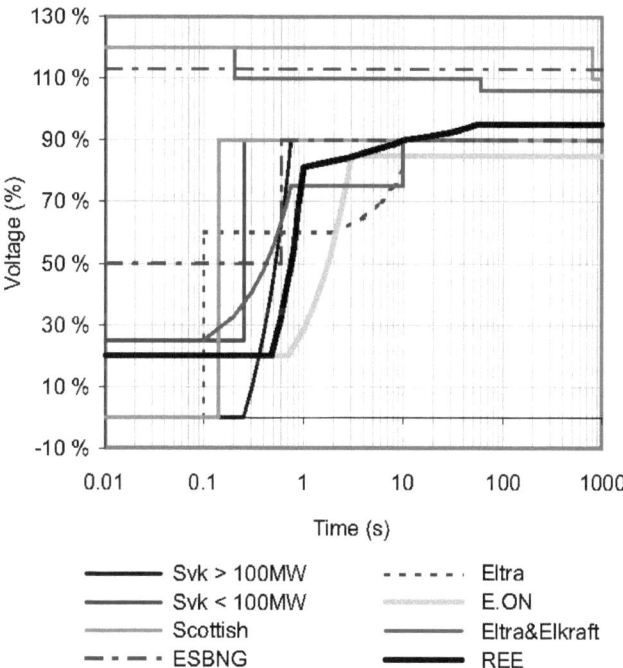

The Spanish grid code[13] establishes that wind farms and all their components must support, without disconnection, voltage dips present at the electric network interconnection point, originated by three-phase, two-phase to ground and single-phase to ground faults. It also specifies:

- Power consumptions are not allowed during the fault and the posterior recovery period. However, it is admitted that some reactive power consumptions during 150 ms time interval after the beginning of the fault and 150 ms after the fault clearance occur. These reactive power consumptions must not be higher than 60% of the wind farm power rated in each network voltage period (20 ms) for three-phase faults, and 40% for single and two-phase faults.
- Wind farms must supply the maximum reactive current to the power system during the fault and recovery time period. Nevertheless, these currents must not be greater than 1.5 times the rated current of the wind farm.

III. PV Plants: Real Data Measurements

A. Data Acquisition System

A power quality analyser – fulfilling IEC 61000-4-30, A-class accuracy, frequency synchronisation and absolute time requirements – has been installed in several Spanish PV plants. This analyser, with a maximum 10 MHz sample rate, is used to capture detailed voltage and current waveforms during voltage dips and clearance of the faults. Additionally, powerful trigger options are available to obtain the entire transient time periods.

This power-quality analyser allows us to measure both DC and AC variables, including environmental parameters such as temperature or radiation levels. The power-quality analyser has also been used to measure voltages and currents at the PV electrical substation level. This power-quality analyser can be linked via WiFi to a UMTS/GPRS modem, offering remote access to the power-quality analyser configuration and its recorded data. This solution can be extended to more power- quality analysers, measuring other inverters or PV plants at different points of the power system. Fig. 2 shows examples of field measurements, where electrical and environmental parameters are collected. In all cases currents and voltages have been measured with a 10.24 kHz sampling frequency.

Fig. 2: PV plant measurements: example

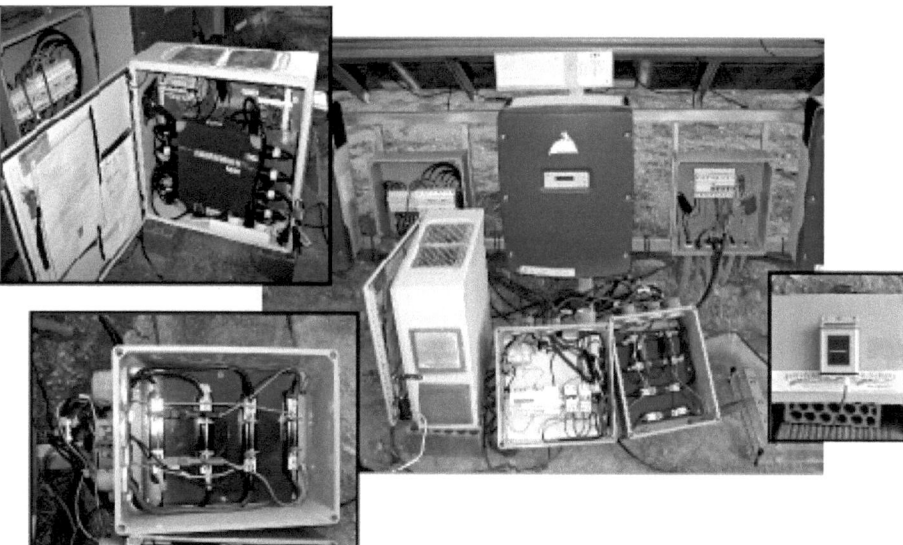

B. Examples of Voltage Dips

Fig 3 shows an example of an unbalanced voltage dip recorded from the grid-side of a three-phase Sunways NT10000 inverter. This inverter is connected in parallel with two more similar inverters, the three SMC 7000HV inverters spread out in different LV lines (around 50 m). In this case, all inverters remain switched on during the disturbance, decreasing the electrical power transmitted to the grid-side during the voltage dip.

Two more unbalanced voltage dips are shown in Fig. 4, corresponding to examples of voltage-dip disturbances measured in PV plants. In this case, the disturbance is considerably shorter than the previous one, being two phases affected by the voltage dip.

Fig. 3: Example of unbalanced voltage dip: three-phase inverter (I)

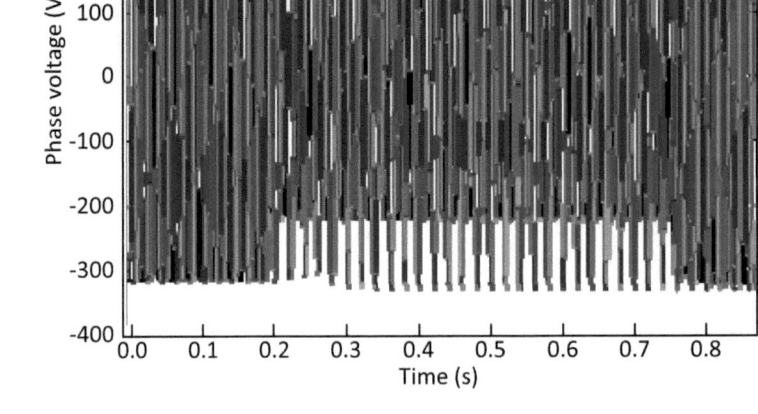

Fig. 4: Example of unbalanced voltage dip: three-phase inverter (II)

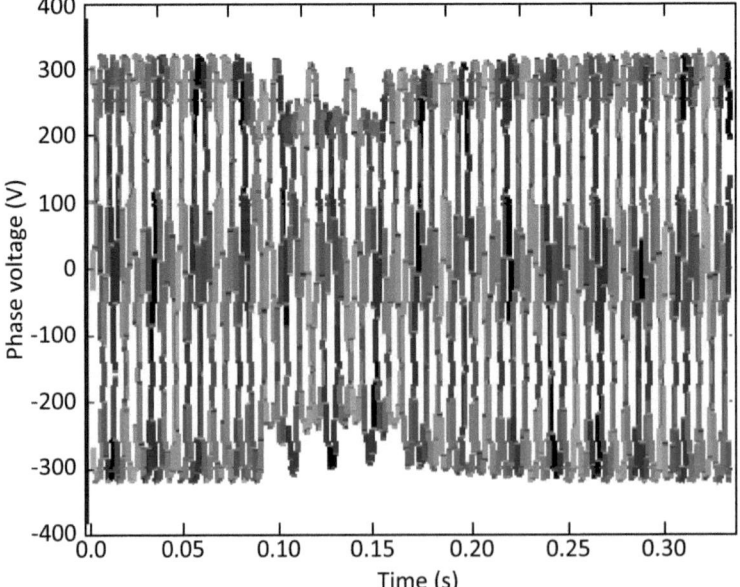

Fig. 5: Example of voltage dip: single-phase inverter (I)

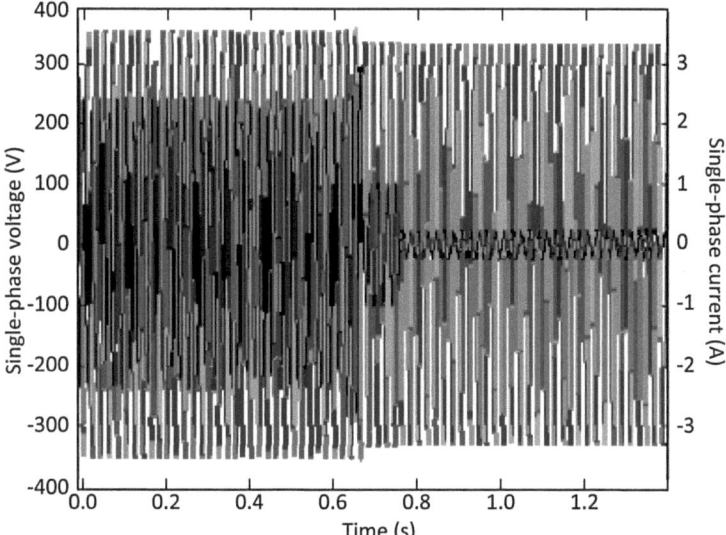

Figure 5 shows a disturbance measured in the grid-side of a single-phase inverter (7000HV). This PV plant only has this single-phase inverter, which was switched off during the voltage dip.

IV. Wind Farms: Real Data Measurements

Data collected in different wind farms under the presence of voltage dips are discussed in this Section. Remarkable unbalanced two-phase and balanced three-phase voltage dips are shown and compared with previous measurements. In this case, two PQ analysers – fulfilling IEC 61000-4-30 class A accuracy, frequency synchronisation, and absolute time requirements – have been installed in a Spanish wind farm, being part of a voltage-dip survey carried out by the authors. These analysers, with a 10.25-kHz sample rate per channel, are able to capture detailed voltage and current waveforms during the voltage dip and the clearance of the fault. Further information can be found in[14], see Fig. 6.

Fig. 7 shows an example of a voltage dip recorded in Spanish wind farms. The voltage dip starts as a phase-to-phase voltage dip, becoming a three-phase voltage dip along the last time interval. This last stage of the dip – around 40% of the dip time – is a balanced voltage dip before recovering the nominal voltage values; i.e. an example of a multi-stage voltage dip.

Fig. 6: Power quality analyser installation scheme

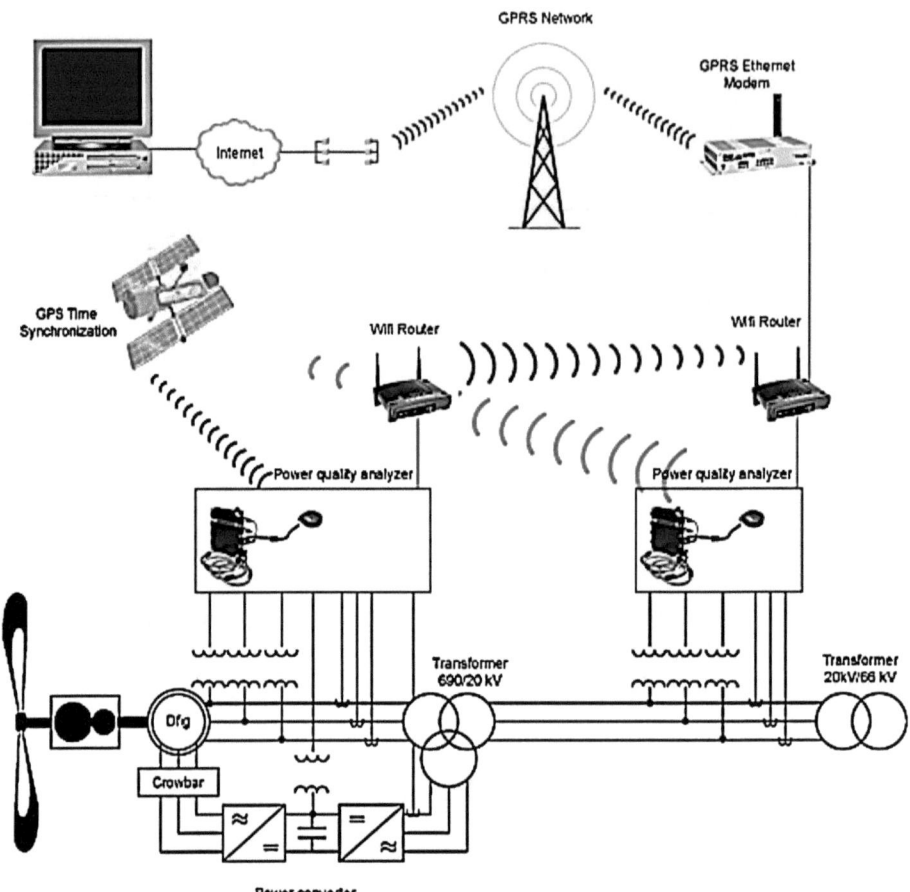

Fig. 7: Example of multistage voltage dip: Wind generator (I)

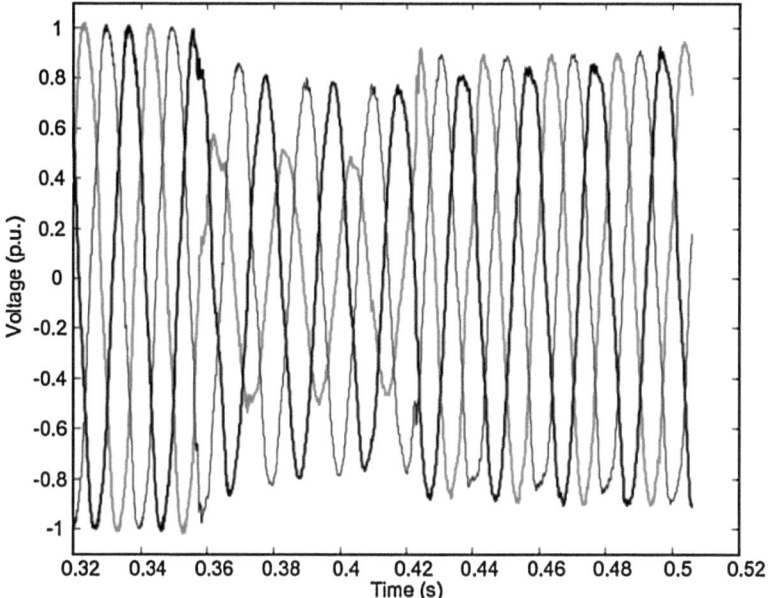

Fig. 8: Example of multistage voltage dip: Wind generator (II)

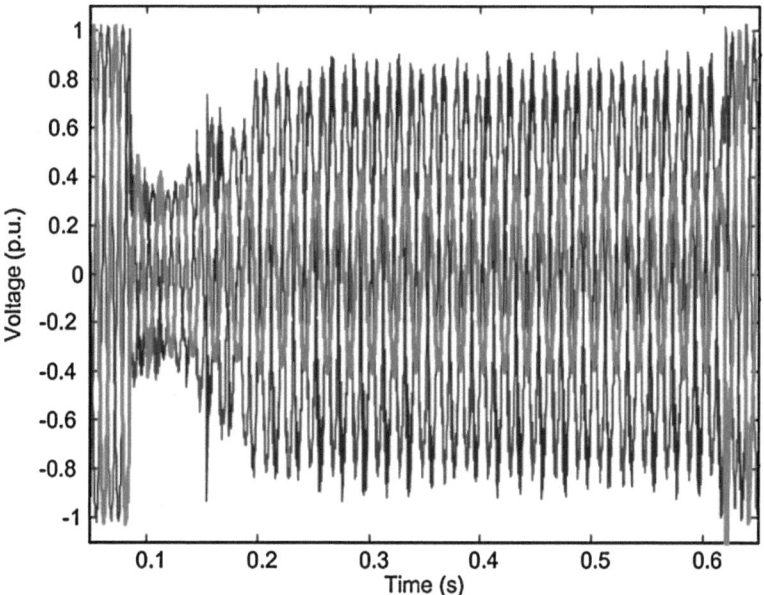

Fig. 8 shows another voltage-dip example collected from the wind turbine nacelle – Gamesa G90 2.0 MW – between the Doubly-Fed Induction Generator (DFIG) and the 0.69/20-kV power transformer. In this case, a multistage disturbance is shown too, where a three-phase voltage dip evolves along the time interval. At the beginning, the three-voltage phases drop significantly in a similar way, and during the rest of the time period all phases are gradually recovered as a symmetrical fault.

From the different examples of real voltage dips previously discussed and taking into account the real data collected in Spanish renewable power plants, the duration and severity of the disturbances is very similar in both wind farm and PV plant measurements. In both cases, power electronic devices are submitted under similar voltage disturbances. In a mid-term, if the grid codes are modified to include PV plants within their requirements, it would thus be necessary to enhance power electronics and keep these renewable power plants connected.

V. Conclusions

A field campaign was carried out in Spanish PV plants to collect real electronic data measured over several months, considering voltage data from both AC and DC sides. These voltage dips in PV plants were measured to offer a preliminary study of the most common disturbances and transients currently suffered from this kind of power generation. These disturbances were compared with previous real data recorded in Spanish wind farms, where the requirements of the international grid codes were clearly more severe. From the results, similar disturbances were found in both renewable power plants as being shorter in duration, as in the case of PV plants. This electronic data gives a preliminary idea about the modifications required in PV plant power electronics when this power generation is included within the international grid codes in a similar way to the wind farms. Additionally, and from the measured data, PV plants offer similar results in grid power quality and voltage harmonic content in comparison with wind farms, using in both cases a power electronic stage as interface between the source of energy and the grid.

VI. Acknowledgment

The authors gratefully acknowledge the contributions of *Gehrlicher Solar Spain* for its work on the original version of this document.

VII. References

[1] "European Standard EN 50160: Voltage characteristics of electricity supplied by public distribution systems".
[2] M.H.J. Bollen, G. Olguin and M. Martins (2005), "Voltage dips at the terminals of wind power installations" Wind Energy, vol. 8, pp. 307-318.
[3] J. Wang, S. Chen and T.T. Lie (2006). "A systematic approach for evaluating economic impact of voltage dips" Electric Power Systems Research.
[4] "Recommended practice for evaluating electric power system compatibility with electronic process equipment," IEEE Std 1346-1998.
[5] E. Gómez, J.A. Fuentes, A. Molina-García, F. Ruz and F. Jiménez (2006). "Results using different reactive power definitions for wind turbines submitted to voltage dips: Application to the spanish grid code", in 2006 IEEE Power Systems Conference and Expo, October-November.
[6] C. Abbey and G. Joos (2005). "Effect of low voltage ride through characteristic on voltage stability", in Power Engineering Society General Meeting, June 2005, pp. 1901-1907.
[7] M.H.J. Bollen, G. Olguin and M. Martins (2004). "Voltage dips at the terminals of wind power installations" in Nordic Wind Power Conference, Goteborg, March 2004.
[8] M.H.J. Bollen (2003), "Algorithms for characterizing measured three-phase unbalanced voltage dips," IEEE Transactions on Power Delivery, vol. 18, no. 3, pp. 937-944, July 2003.
[9] M.H.J. Bollen and L.D. Zhang (2003). "Different methods for classification of three-phase unbalanced voltage dips due to faults" Electric Power Systems Research, vol. 66, pp. 59-69.
[10] L. Zhang and M.H.J. Bollen (2000). "Characteristic of voltage dips (sags) in power systems" IEEE Transactions on Power Delivery, vol. 15, no. 2, pp. 827-832, April 2000.
[11] R. Leborgne and D. Karlsson (2005). "Phasor based voltage sag monitoring and characterisation" in International Conference on Electricity Distribution, June 2005.
[12] J. Matevosyan, T. Ackermann, S. Bolik and L. Söder (2004). "Comparison of international regulations for connection of wind turbines to the network" in Nordic wind power conference, Gothenburg, March 2004.
[13] "Response requirements in front of voltage dips at wind farms utilities". Spanish MITC. BOE nº 254, October 24, 2006.
[14] E. Gómez-Lázaro, J.A. Fuentes, A. Molina-García and M. Cañas-Carretón (2009). Characterization and Visualization of Voltage Dips in Wind Power Installations, IEEE Trans on Power Delivery, vol. 24, no. 4, pp. 2071-2078, October 2009.

Renewable Energy Policies that Impact Climate Change

–

The Case for Photovoltaic Solar Technology

Nasir J. Sheikh, Tugrul U. Daim

Abstract

The purpose of this paper is to provide an overview of the global status and trends of renewable energy and the policy targets and landscape driving these trends. Typically energy trends and polices do not include guidance or mandate for long-term technology development policies or technology production policies. With the aid of literature review the authors attempt to provide a more encompassing view of the energy policies with a focus on photovoltaic solar technology (PVST) wherever possible. These policies tend to be broad-based government mandates covering multiple types of renewable and it is not always possible to identify specific PV policies. In fact government policies purposely avoid preference of one type of technology or approach. Instead broad targets are defined coupled with incentives (also referred to as "promotion policies") to achieve these targets.

I. Introduction

It is well established that adoption of renewable energy (commonly referred to as "renewables") is an important factor for the mitigation of greenhouse gases (GHG) and their impacts on climate change. [Other factors include: innovation and new technologies, Cap and Trade Policy, energy policies, Smart Grids, etc.] There are multiple renewable energy sources and formats including: solar, wind, geothermal, biomass, and hydro (or hydropower).

In the past five years photovoltaic (PV) solar energy has shown the highest growth increase at six-fold rate[1]. Hence the adoption of photovoltaic solar technology (PVST) is of special interest. Energy policies are a major driver (or barrier) to the adoption of renewables, especially PV.

To fulfill the stated purpose the following research questions are to be answered:

- What is the current global policy landscape for PV (and renewables, in general)?
- What is the impact of policy on innovation?
- What type of technology policies are needed for long-term climate change?
- What is the energy viability of PV systems and how does that impact policy?

II. Research Method

Fig. 1: Research Methodology

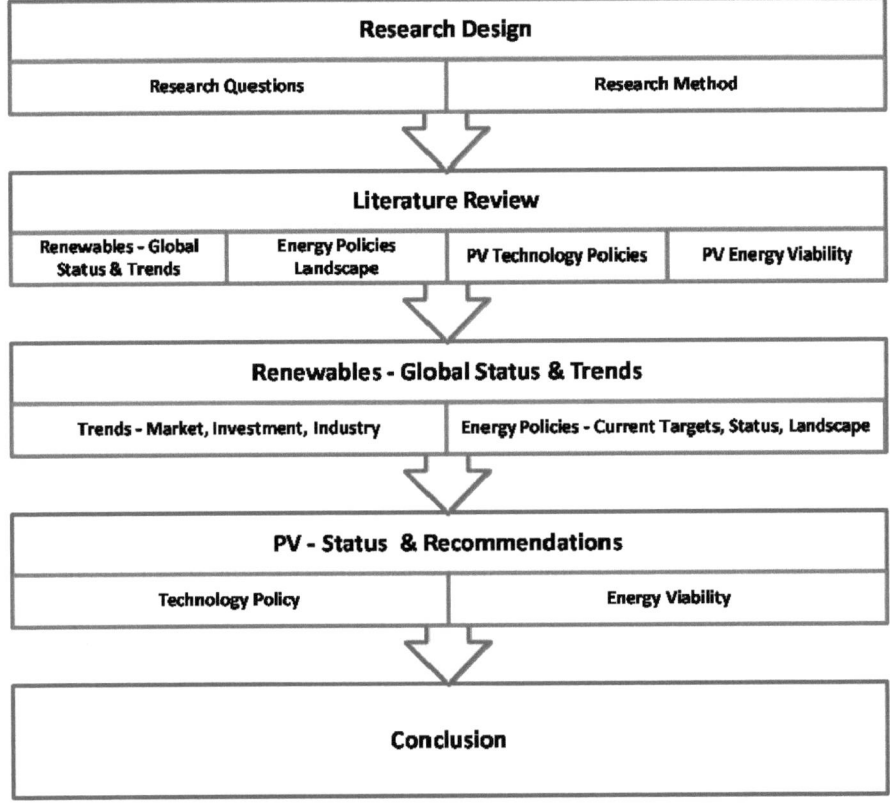

The research method utilized is essentially a literature survey and analysis with a view to present value of the current established policies combined with recommendations for future policies to cover long-term PV technology innovation and energy viability (Fig. 1). In an effort to address the research questions as stated above, the research method starts with a literature review of global status, trends, and policies in renewable and extends to technology policies and energy viability of PV. Then with a focus on PV the view of domain experts is presented. Finally, this research concludes with PV specific policies to be included in renewable policies to mitigate climate change.

III. Renewable Energy – Global Status and Trends

The global status and trends of renewables is presented in the following way:
- Global Status
- Photovoltaic Market Overview
- Investment Trends
- Industry Trends
- Policy Landscape.

A. Global Status

Since 2004 many indicators of renewable energy have shown dramatic gains[1, 2]. Investments in renewable energy have increased by four times to become $120 billion in 2008 (Table 1). In the same period solar photovoltaic (PV) capacity increased six times to exceed 16 gigawatts (GW), wind energy capacity increased 3.5 times to 121 GW, and total power capacity from new renewables increased 75% to 280 GW. This included gains in small hydro, geothermal, and biomass power. Solar heating capacity doubled to 145 gigawatts-thermal (GWth), biodiesel production increased six times to 12 billion liters per year, and ethanol production doubled to 67 billion liters per year.

During 2008 annual percentage increases were even more outstanding. Wind power grew by 29%, grid-connected solar PV by 70%, utility-scale solar PV plants (larger than 200 kilowatts) tripled to 3 GW, solar hot water grew by 15%, ethanol and biodiesel production both grew by 34%, and small hydroelectric power grew by 8%. Biomass and geothermal heat and power also continued to increase.

In fact, 2008 was a record year for changes affecting renewable energy leadership, markets, and policy. The United States overtook other countries with $24 billion invested in new capacity (and this was 20% of global total investment) (Table 2). The United States also led in new wind power capacity. Spain added 2.6 GW of solar PV which was 50% of the total global grid-connected installations. China doubled its wind power and became the fourth largest producer of wind-generated power. Also for the first time, the United States and the European Union added more power capacity from renewables than from traditional sources such as gas, coal, oil, and nuclear.

Renewable energy industries boomed during 2008. In 2008 global solar PV production increased by 90% to 6.9 GW. China overtook Japan to become the new world leader in PV cell production and also experienced high growth in its wind power industry, with multiple new companies producing wind turbines and components. Globally, the wind industry continued to push turbine sizes higher, with models of 3 MW or larger becoming more widespread. The concentrating solar power (CSP) industry saw many entrants and new manufacturing facilities. The ethanol and biodiesel industries similarly expanded, particularly in North America and Latin America, and the cellulosic ethanol industry was in the process of booming, with 300 million liters per year of capacity under construction.

Although the clean energy sector initially fared better than many other sectors in the financial crisis in late, renewable investment did experience a downturn after September 2008. However, the number of projects continued to increase and many economic stimulus bills included provisions for renewable energy. In developing countries investments in renewables increased to about $2 billion in 2008.

By early 2009, policy targets existed in at least 73 countries, and at least 64 countries had policies to promote renewable power generation, including 45 countries and 18 states/provinces/territories with feed-in tariffs (many of these recently updated). The number of countries/states/provinces with renewable portfolio standards increased to 49. Policy targets for renewable energy were added or upwardly revised in a large number of countries in 2008. Many forms of policy support for renewables were also introduced or extended. For example, governments of Australia, China, Japan, Luxembourg, the Netherlands, and the United States provided new solar PV subsidies; new laws and policy provisions for renewables were enacted in developing countries, including Brazil, Chile, Egypt, Mexico, the Philippines, South Africa, Syria, and Uganda; new mandates for solar hot water and other renewable heating were introduced in Cape Town (South Africa), Baden-Württemberg (Germany), Hawaii, Norway, and Poland; new biofuels blending mandates or targets appeared in at least 11 countries with India stating a a new 20% target; and the number of global green power consumers grew to 5 million households and businesses. City and local government policies grew segments of the policy landscape, with several hundred local or municipal governments planning or implementing renewable energy policies. Planning frameworks linked to reduction of carbon dioxide emissions were also part of this new wave.

The PV market proved to be a front-runner in renewables and is discussed in greater detail in the next section.

Table 1: Selected Indicators[1]

SELECTED INDICATORS	2006	2007	2008
Investment in new renewable capacity (annual)	63	104	120 Billion USD
Renewables power capacity (existing, excluding. large hydro)	207	240	280 GW
Renewables power capacity (existing, including large hydro)	1020	1,070	1,140 GW
Wind power capacity (existing)	74	94	121 GW
Grid-connected solar PV capacity (existing)	5.1	7.5	13 GW
Solar PV production (annual)	2.5	3.7	6.9 GW
Solar hot water capacity (existing)	105	126	145 GWth
Ethanol production (annual)	39	50	67 Billion Liters
Biodiesel production (annual)	6	9	12 Billion Liters
Countries with policy targets		66	73
States/provinces/countries with feed-in policies*		49	63
States/provinces/countries with RPS policies*		44	49
States/provinces/countries with biofuels mandates*		53	55

*Notes: *A feed-in tariff (FIT) is an energy-supply policy to encourage new renewable electricity generation. In the United States, FIT policies may require utilities to purchase either electricity, or both electricity and the renewable energy (RE) attributes from eligible renewable energy generators. The FIT contract provides a guarantee of payments in dollars per kilowatt hour ($/kWh) for the full output of the system for a guaranteed period of time (typically 15-20 years). In Europe, FIT policies may or may not include the attributes. It is presumed that under current US law that payment for the power would be made under Federal Energy Regulatory Commission (FERC) wholesale power rules, and payment for the RECs could be made under state law. However, this is an assumption, and these issues will need to be clarified using a proper legal review in due course. RPS policies require electric utilities to provide renewable electricity to their customers, typically as a percentage of total energy use. Twenty-eight states and the District of Columbia have mandatory RPS policies, five states have voluntary RPS goals (DSIRE 2009c), and more states (as well as the federal government) are considering implementing similar policies. In Europe, RPS policies are called quota-based mechanisms, quota obligations, or renewables obligations*[3].

Table 2: Top Five Countries[1]

TOP FIVE COUNTRIES	#1	#2	#3	#4	#5
Annual Amounts - 2008					
New capacity investment1	United States	Spain	China	Germany	Brazil
Wind power added	United States	China	India	Germany	Spain
Solar PV added (grid-connected)3	Spain	Germany	United States		
			South Korea		
			Japan		
			Italy		
Solar hot water/heat added4	China	Turkey	Germany	Brazil	France
Ethanol production	United States	Brazil	China	France	Canada
Biodiesel production	Germany	United States	France	Argentina	Brazil

B. Photovoltaic Market Overview

Renewable energy markets grew robustly in 2008 (Fig. 2)[1], [3]. Among new renewables (excluding large hydropower), wind power was the largest addition to renewable energy capacity. However, grid-connected solar photovoltaic (PV) continued to be the fastest growing power generation technology with a 70% increase in existing capacity to 13 GW in 2008. This represents a six fold increase in global capacity since 2004 (Fig. 3). Annual installations of grid-tied solar PV reached an estimated 5.4 GW in 2008. Spain led other countries with 2.6 GW of new capacity which was an increase of five times over 2007 and representing half of new global installations. Spain's unprecedented increase surpassed former PV leader Germany which had installed 1.5 GW in 2008. Other major markets in 2008 included the United States, South Korea, Japan, and Italy which added 310 MW, 250 MW, 240 MW, and 250 MW respectively. Other major markets such as Australia, Canada, China, France, and India also grew followed by emerging markets such as China. Total global installed PV capacity increased to 16 GW in 2008.

Solar PV markets showed three clear trends in 2008. The first was the growing attention to building-integrated PV (BIPV), which is a small but fast-growing segment of some markets, with more than 25 MW installed in Europe. Second, thin-film solar PV technologies became a larger share of total installations. And third, utility-scale solar PV power plants (defined as larger than 200 kilowatts, kW) emerged in large numbers in 2008. By the end of 2008, an estimated 1,800 such plants existed worldwide, up from 1,000 at the end of 2007. Altogether, these plants totaled over 3 GW, a tripling of existing capacity from 2007. These power plants were added mainly in Spain (over 1.9 GW added) followed by the Czech Republic, France, Germany, Italy, Korea, and Portugal. The 60-MW Olmedilla de Alarcon plant in Spain became the world's largest solar PV plant. New plants are under development in Europe, China, India, Japan, and the United States.

Fig. 2: Solar PV, Existing World Capacity, 1995-2008[1]

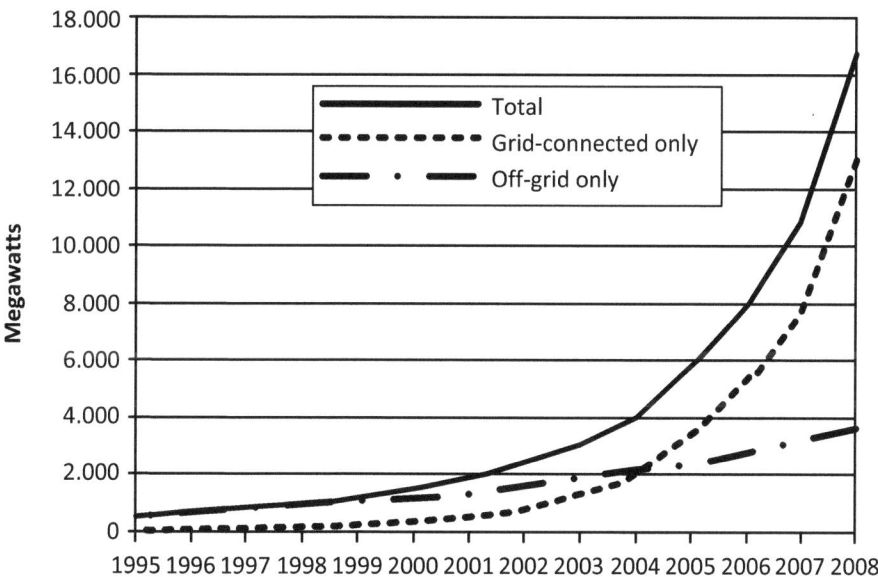

Fig. 3: Renewable Power Capacities, Developing World, EU and Top Six Countries, 2008[1]

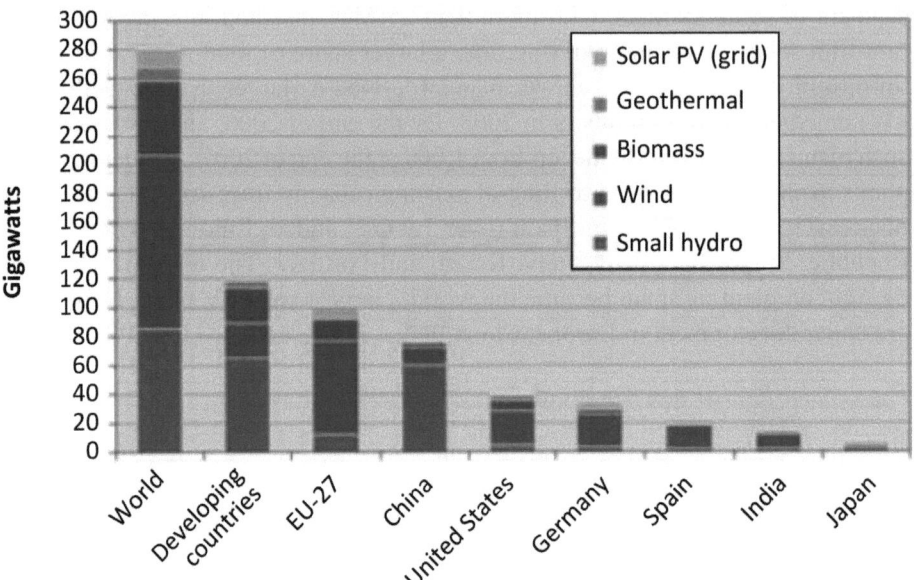

C. Investment Trends

Global renewable energy investments included $120 billion for new capacity and biofuel plants (Fig. 4–Fig. 9) doubling from 2006, with the increase being mainly from wind energy, solar, and biofuels[4]. The 2008 investment breakdown was wind energy – 42%, solar PV – 32%, biofuels – 13%, biomass and geothermal – 6%, solar hot water – 6%, and small hydroelectric power – 5%. (Additionally, $45 billion was spent on large hydroelectric power.)

In terms of new investments and technology incubation solar PV far exceeded other renewable (Fig. 9).

In 2008, the United States led investments with $24 billion (which was 20% of total global investments) due to record wind energy installations and ethanol investments beating out Germany. Spain, China, and Germany followed closely, with $15-19 billion each. Brazil was next, at $5 billion, mainly due to biofuels. Additionally, the solar PV and wind industries invested in new manufacturing plant with global research and development (R&D) alone exceeding $15 billion. Large private equity and venture capital flows (exceeding $13.5 billion in 2008) continued until the market crash of late-2008.

At the end of 2008 and in early 2009, in part in response to the financial crisis, a number of national governments announced plans to greatly increase public finance of renewable energy and other low-carbon or clean technologies. Many of these announcements were directed at economic stimulus and job creation, with millions of new "green jobs" targeted. The United States led with a target of $150 billion for renewables over 10 years, Japan planned to spend 1 trillion yen ($12.2 billion) over five years, Hungary is investing €250 million ($330 million) over seven years, South Korea announced a $36 billion package over four years, and Australia planned to accelerate an existing AUD$500 million ($370 million) renewable energy fund from the original six years to one and half years. The United States also authorized $1.6 billion in tax credit bonds – a significant increase from previous years – to finance renewables investments. The Netherlands set aside €160 million ($200 million) per year for 15 years to support offshore wind power. China has been providing increasing amounts of public financing to renewables – about $300 million for one period in 2007/2008, and at the end of 2008 it announced $15 billion support for renewable energy, mainly for wind energy. Mexico's new 2008 renewable energy law established an $800 million fund for renewable energy projects. Morocco pledged a $1 billion fund for renewables and energy efficiency programs.

Fig. 4: Global Investment in Renewable Energy, 2004–2008[4]

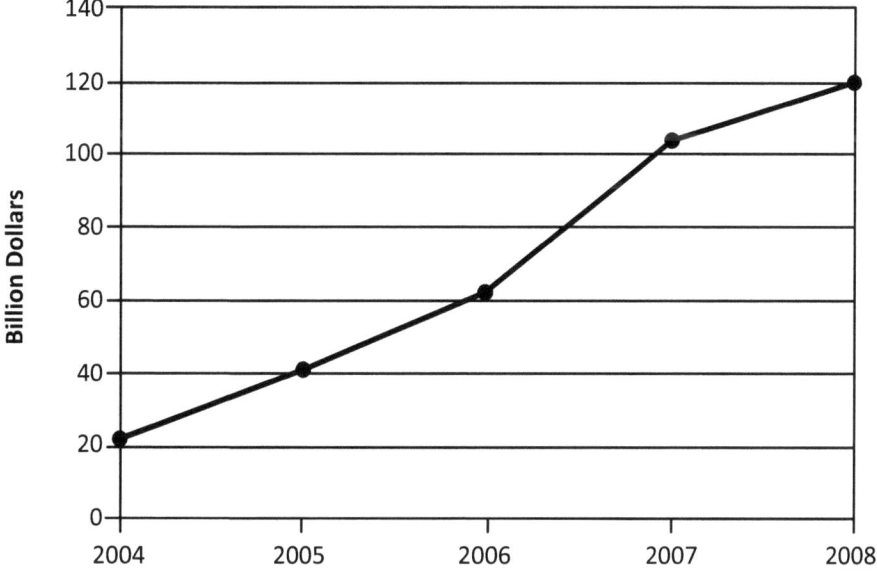

Fig. 5: Global New Investment by Technology, 2007[4]

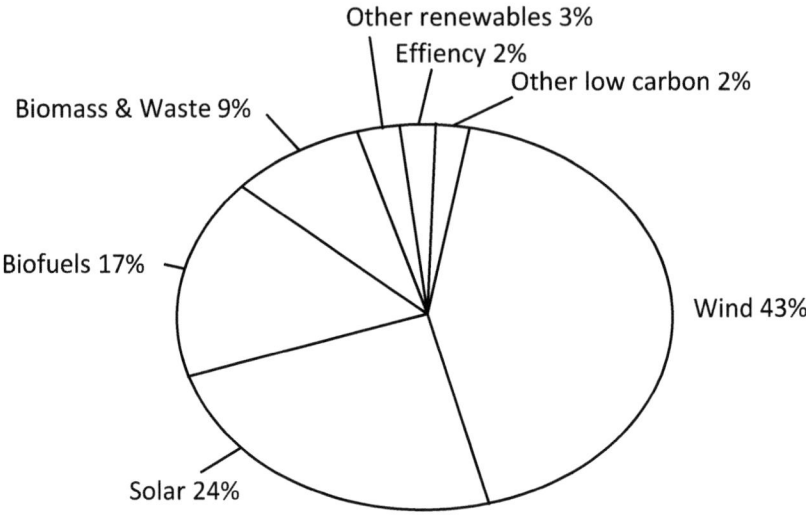

Fig. 6: Asset Finance New Investment by Technology[4]

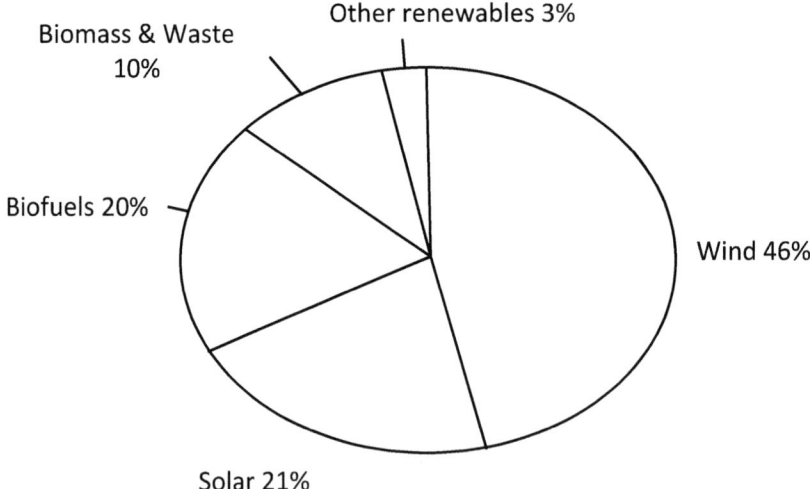

Fig. 7: Public Markets New Investment by Technology, 2007[4]

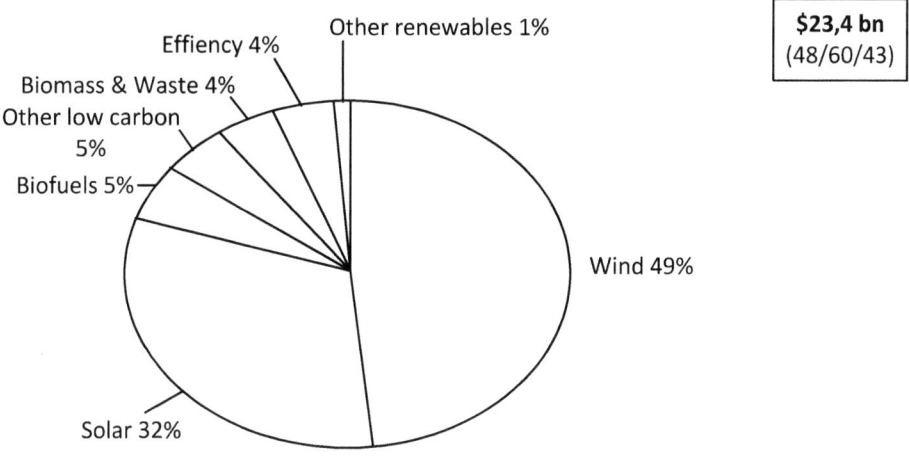

Fig. 8: Venture Capital / Private Equity (VC/PE) New Investment by Technology, 2007[4]

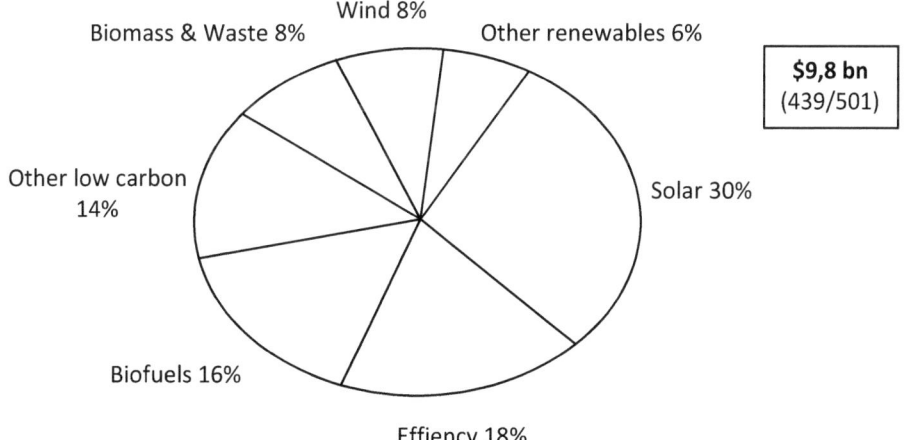

Fig. 9: Clean Energy Incubatees by Sector, 2007[4]

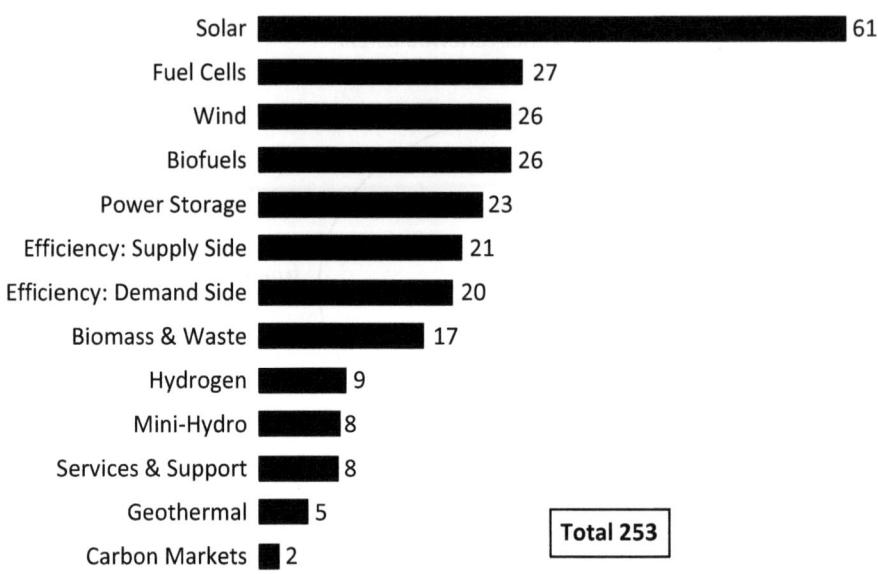

D. Industry Trends

In 2008, the renewable energy industry grew significantly in manufacturing capacity and there were shifts in leadership[1]. Publicly traded renewable energy companies worldwide had a market capitalization of $240 million (prior to the market crash of 2008). The number of companies in this category also increased to 160 from 60 in 2005. Examples of such companies included wind developers Iberdrola Renovables and EDP Renovavies, wind turbine transmission equipment manufacturer – Hansen Transmission, and solar ingots and wafers producer – Crystallox. However, after the 2008 market crash all companies took a hit and some had to close plants and lay off workers. However, many companies continued to do well in early 2009[6].

The solar PV industry continued to be one of the world's fastest growing industries. During 2004 and 2008, the global annual production increased sixfold to 6.9 GW and in 2008 annual production was 90% higher than in 2007. China became the new world leader in PV cell production (1.8 GW) by overtaking Japan, the previous leader. Germany was second (1.3 GW), followed by Japan (1.2 GW), Taiwan (0.9 GW), and the United States (0.4 GW). The United States ranked fifth in PV production but ranked first in thin-film production (270 MW), followed by

Malaysia (240 MW) and Germany (220 MW). Global thin-film production increased 120% in 2008, to reach 950 MW[7].

Market share of companies also changed in a mjor way during 2007-2008. Q-Cells of Germany was the top global PV manufacturer in 2007 and 2008 (with 570 MW in 2008). First Solar doubled its production in 2008 to 500 MW. Suntech of China tripled its production to 500 MW, tying with Suntech, and later claimed a further increase in capacity to 1 GW. Sharp of Japan fell to fourth place (from first place), with 470 MW in 2008[8].

In 2008 the PV industry had 8 GW of cell manufacturing capacity which included 1 GW of thin-film. The industry also had plans for major capacity expansion in thin-film. However, the market crash of 2008 forced many companies to revise their expansion plans. During 2008 India rose to prominence as the new producer of solar PV supported by national and state government policies for PV manufacturing in special economic zones with investment subsidies of 20% resulting in new $18 billion investment plans[9].

E. Policy Landscape

1. Policy Targets

Renewable energy policies existed in the 1980s but only within a few countries. During 1998-2008 decade these policies started to take hold in many more countries, local regions, and cities. The pace has accelerated in the last five years. Many of these policies have exerted substantial influence on the market development reviewed in the previous sections. This section first covers existing policy targets for renewables, and then reviews policies to promote renewables. It also discusses green power purchasing and certificates and municipal-level policies.

In 2008 many countries added or revised renewable energy policy targets and by 2009, policy targets for renewable energy existed in at least 73 countries. This included state or provincial-level targets in the United States and Canada, which have no national targets. In addition, an EU-wide target was enacted in 2007[1] (Tables 3–7). Twenty-nine U.S. states (and the District of Columbia) and nine Canadian provinces have targets based on renewables portfolio standards (even though the United States and Canada do not have national targets). National targets

---

1 The term "target" is used as a general term to broadly cover multiple types of policies, legislative mandates, government decisions, programs, or pledges made as part of international joint-action programs (from Bonn Renewables 2004, Beijing International Renewable Energy Conference 2005, and Washington International Renewable Energy Conference 2008 (WIREC))[3].

are typically for shares of electricity production, around 5%-30%, and may range from 2% to 78%. Other targets are for shares of total primary energy supply, installed capacity, or renewable energy production. The timeframes being considered are for 2010-2012, 2020, or 2025. In 2007, the European Commission agreed to new binding targets for 2020 of 20% of final energy (implying 34% of electricity would be provided by renewable) (Fig. 10). These new targets extend the existing targets of 21% of electricity and 12% of primary energy by 2010. Similar to the existing electricity targets, individual countries will need to adopt their own national targets to meet the European Commission's 20% target (Table 3). Several countries like Netherlands and Germany have already adopted a 20% and 25%-30% target of final energy respectively. Germany plans to increase its target further by 2030 up to 45%.

Fig. 10: EU Renewable Energy Targets – Share of Final Energy by 2020

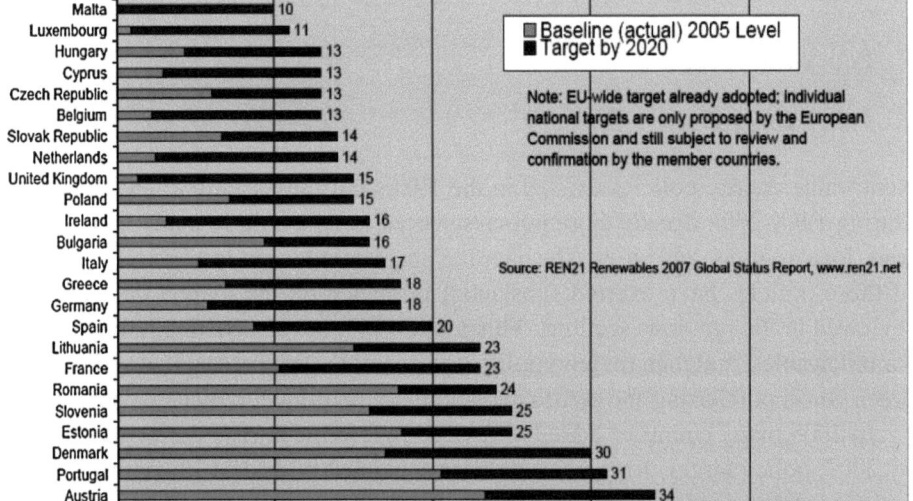

The countries with national targets include 22 developing countries: Algeria, Argentina, Brazil, China, the Dominican Republic, Egypt, India, Indonesia, Iran, Jordan, Malaysia, Mali, Morocco, Nigeria, Pakistan, the Philippines, Senegal, South Africa, Syria, Thailand, Tunisia, and Uganda.

Among developing countries, China received considerable attention when it confirmed its targets in its new long-term renewables development plan, issued in September 2007. China's national target is 15% of primary energy by 2020, and there are individual technology targets as well, including 300 GW of hydro, 30 GW of wind, 30 GW of biomass, and 1.8 GW of solar PV. Meeting these targets would almost triple China's renewable energy capacity by 2020 (Table 4).

Table 3: Share of Primary and Final Energy from Renewables, Existing in 2006 and Targets[3]

	Primary energy (IEA method)		Final energy (EC method)	
Country/region	Existing share (2006)	Future target	Existing share (2005–06)	Future target
World	13%	—	18%	—
EU-25/EU-27	6.5%	12% by 2010	8.5%	20% by 2020
Selected EU Countries				
Austria	20%	—	23%	34% by 2020
Czech Republic	4.1%	8–10% by 2020	6.1%	13% by 2020
Denmark	15%	30% by 2025	17%	30% by 2020
France	6.0%	7% by 2010	10%	23% by 2020
Germany	5.6%	4% by 2010	5.8%	18% by 2020
Italy	6.5%	—	5.2%	17% by 2020
Latvia	36%	6% by 2010	35%	42% by 2020
Lithuania	8.8%	12% by 2010	15%	23% by 2020
Netherlands	2.7%	—	2.4%	14% by 2020
Poland	4.6%	14% by 2020	7.2%	15% by 2020
Spain	6.5%	12.1% by 2010	8.7%	20% by 2020
Sweden	28%	—	40%	49% by 2020
United Kingdom	1.7%	—	1.3%	15% by 2020
Other Developed/OECD Countries				
Canada	16%	—	20%	—
Japan	3.2%	—	3.2%	—
Korea	0.5%	5% by 2011	0.6%	—
Mexico	9.4%	—	9.3%	—
United States	4.8%	—	5.3%	—
Developing Countries				
Argentina	8.2%	—	—	—
Brazil	43%	—	—	—
China*	8%	15% by 2020	—	—
Egypt	4.2%	14% by 2020	—	—
India	31%	—	—	—
Indonesia	3%	15% by 2025	—	—
Jordan	1.1%	10% by 2020	—	—
Kenya	81%	—	—	—
Mali	—	15% by 2020	—	—
Morocco*	4.3%	10% by 2010	—	—
Senegal	40%	15% by 2025	—	—
South Africa	11%	—	—	—
Thailand*	4%	8% by 2011	—	—

Table 4: China Renewable Energy Targets[3]

	2006 actual	2010 target	2020 target
Hydro power	130 GW	190 GW	300 GW
Wind power	2.6 GW	5 GW	30 GW
Biomass power	2.0 GW	5.5 GW	30 GW
Solar PV	0.08 GW	0.3 GW	1.8 GW
Solar hot water	100 million m²	150 million m²	300 million m²
Ethanol	1 million tons	2 million tons	10 million tons
Biodiesel	0.05 million tons	0.2 million tons	2 million tons
Biomass pellets	~ 0	1 million tons	50 million tons
Biogas and biomass gasification	8 million m³/year	19 billion m³/year	44 million m³/year
Share of primary energy	8%	10%	15%

Besides China, several other developing countries adopted or upgraded targets during 2006/2007. Argentina set a target of 8% of electricity from renewables by 2016 (excluding large hydro). Egypt revised its target to 20% share of electricity by 2020, up from the previous target of 14% (which included 7% from hydro). The provincial government of the Western Cape in South Africa set a target of 15% of electricity by 2014. Morocco was drafting a new renewable energy law that would target a 10% share of primary energy and 20% share of electricity by 2012, which would imply 1 GW of new renewables capacity. And Uganda enacted a comprehensive set of targets through 2017 in a new 2007 renewable energy strategy. A number of other developing countries were working on targets expected in the near future, including a proposed "Brasilia Platform on Renewable Energies" by a group of 21 Latin American and Caribbean countries for 10% of primary energy from renewables. Mexico is considering a target of 8% of electricity by 2012, excluding large hydro. And India has proposed long-term goals by 2032 in several categories, including 15% of power capacity.

New or expanded solar PV promotion programs continued to appear around the world at the national, state/provincial, and local levels. Most significant was the opening of the grid-connected solar PV market in China with a new policy for building-integrated PV (solar panels used as architectural components), which also applies to off-grid applications. That policy provides initial subsidies in 2009 of 20 RMB per watt ($3 per watt) for installations larger than 50 kW. (Such minimum capacity caps are unusual globally, as most other subsidy policies set maximum capacity caps.) It also specifies minimum solar cell efficiencies and gives priority to building-integrated systems and public buildings. Some other examples of new and modified solar PV promotion policies include: Japan increased national solar PV subsidies for schools, hospitals, and railway stations from 33% to 50%, in addition to reinstating subsidies for households that had expired in 2005 (although at a lower level of about 10%). Japan also plans to have more than two-thirds of newly built houses equipped with solar PV by 2020. Australia, Luxembourg, and the Netherlands all enacted new solar PV subsidy programs. In the United States, Massachusetts and New Jersey adopted capital subsidies ($1.75 per watt for residential up to 10 kW and $1 per watt for non-residential up to 50 kW in New Jersey). A number of U.S. states and cities have been considering residential solar lease/loan programs, following the lead of emerging programs in Connecticut and the city of Berkeley, California. India enacted a comprehensive support program for solar PV with subsidies and loans to supplement the new national feed-in tariff for solar PV. And Mexico established a standard contract for net metering that includes commercial solar PV installations up to 30 kW.

Many other countries boosted or extended various forms of policy support in 2008 and early 2009. Examples include: Portugal simplified licensing for small renewables producers. Denmark began public investment in wind farms. The province of Ontario in Canada is finalizing a "Green Energy Act" that gives priority to all forms of renewable energy, promotes community involvement, and establishes a "renewable energy czar" at the provincial level. In the United States, the federal 30% investment tax credit (ITC) was extended through 2016 for solar PV, solar thermal power, solar hot water, small wind, and geothermal. The $2,000 cap on residential credits was removed, and ITC eligibility was broadened to include utility companies. And the U.S. production tax credit (PTC) was extended for wind power through 2012, and for biomass, geothermal, hydropower and marine power through 2013. At the state level, net-metering policies were broadened or added, with the result that 44 U.S. states now have some type of net-metering policy. And many net-metering laws continued to increase capacity limits; for example, the province of Nova Scotia in Canada increased the capacity limit from 100 kW to 1 MW.39 Among developing countries, Mexico adopted a new renewable energy law that mandates utility purchases of renewables generation, sets up a project fund, and mandates a national target (to be determined). The Philippines similarly adopted a milestone renewable energy law, which mandates both renewable portfolio standards (to be developed within one year) and feed-in tariffs for wind, solar, biomass, small hydro, and ocean power. The Philippines law also provides connection priority and transmission priority for renewable generators, allows consumers to voluntarily choose to purchase renewable power from suppliers, and provides tax and import-duty incentives for investment. China changed VAT- and import-duty-related promotion mechanisms to further favor domestic wind turbine production. India provided accelerated tax depreciation along with its national feed-in tariff. Egypt's new electricity law enables independent renewable power producers and calls for feed-in tariffs. Syria's new

Energy Conservation Law encourages private-sector renewable power generation. Uganda increased an off-grid solar PV capital subsidy from 14% to 45%. South Africa created a new public agency to accelerate renewables projects. And Chile's new 2008 national renewable energy development program created a market-facilitation, best-practice, and promotion center for renewables.

2. City and Local Government Policies

Policies for PV and renewable energy are a diverse and growing segment of the renewable energy policy landscape, with several hundred cities and other forms of local government around the world adopting goals, promotion policies, urban planning, demonstrations, and many other activities. (Some of these are highlighted in Table 5 and Table 6).

Table 5: Selected Cities with Renewable Energy Goals/Policies[1], [3]

City	Renewable energy goals	CO_2 reduction goals	Policies for for solar hot water	Policies for solar PV	Urban planning, pilots, and other policies
Adelaide, Australia	✓	✓			✓
Austin (Texas), USA	✓	✓			✓
Barcelona, Spain			✓		
Berlin, Germany		✓	✓	✓	
Betim, Brazil		✓	✓		✓
Cape Town, South Africa	✓	✓			✓
Chicago, USA	✓				
Daegu, Korea	✓	✓			✓
Freiburg, Germany	✓	✓	✓	✓	✓
Gwangju, Korea	✓	✓			✓
The Hague, Netherlands		✓			
Leicester, UK	✓				✓
London, UK		✓			
Malmö, Sweden		✓			✓
Melbourne, Australia	✓	✓			✓
Mexico City, Mexico				✓	✓
Minneapolis, USA	✓				✓
Nagpur, India		✓	✓	✓	
New York, USA		✓		✓	✓
Oxford, UK	✓	✓	✓	✓	✓
Portland, United States	✓	✓	✓	✓	✓
Rizhao, China			✓	✓	
Salt Lake City, USA	✓	✓			✓
Santa Monica, USA	✓				✓
São Paulo, Brazil			✓		
Sapporo, Japan		✓			✓
Stockholm, Sweden	✓	✓			✓
Toronto, Canada		✓			
Tokyo, Japan	✓		✓	✓	✓
Townsville, Australia			✓	✓	
Vancouver, Canada		✓			
Växjö, Sweden	✓	✓	✓	✓	✓
Woking, UK	✓	✓	✓	✓	✓

Table 6: Selected Municipal Targets and Goals for Renewable Energy[1], [3]

City	Targets for renewable share of electricity	CO2 emissions reductions goals	Other targets/goals
Austin (TX), USA	30% by 2020	carbon-neutral by 2020	100% of own elec. use by 2012
Adelaide, Australia	15% by 2014	transport/buildings zero net emissions by 2010/12	2 MW of solar PV on residential and commercial buildings
Berlin, Germany		25% below 1990 by 2010	
Cape Town, South Africa	10% by 2020		10% of homes by 2010 with solar hot water
Chicago, USA			20% of own elec. use by 2006
Daegu, Korea			5% of energy by 2012
Freiburg, Germany	10% by 2010	25% below 1992 by 2010	
Gwangju, Korea		20% below 1990 by 2020	2% of energy by 2020
Leicester, UK			10% of energy by 2010 and 20% by 2020
London, UK		20% below 1990 by 2010	
Malmö, Sweden		25% below 1990 by 2012	
Melbourne		20% below 1996 by 2010	25% RE in buildings by 2010
New York, USA		7% below 1990 by 2012	
Oxford, UK			10% of homes by 2010 with solar hot water/PV
Portland (OR) USA		10% below 1990 by 2010	100% of own elec. use by 2010
Sacramento, USA	20% by 2011		
Salt Lake City, USA			10% of new building energy use
San Francisco, USA			1 MW/year added
Santa Monica, USA			100% of own use (current)
Sapporo, Japan		10% below 1990 by 2012	
Tokyo, Japan			20% of energy by 2020 (proposed); 5% of own use
Toronto, Canada		30% by 2020; 80% by 2050	
Vancouver, Canada		30% by 2020; 80% by 2050	

3. Renewables Promotion Policies

At least 60 countries – 37 developed and transition countries and 23 developing countries – have some type of policy to promote renewable power generation (Table 7)[1]. The most common policy is the Feed-in Law, which has been enacted in many new countries and regions in recent years. A Feed-in Law or Tariff is an incentive structure to encourage the adoption of *renewable energy* through government *legislation*[4]. The regional or national *electricity utilities* are mandated to buy *renewable electricity* (*electricity* generated from *renewable sources*, such as *solar photovoltaics*, *wind power*, *biomass*, *hydropower* and *geothermal power*) at above-market rates set by the government. In 1978 the United States was the first country to enact a national feed-in law. Feed-in policies were next adopted in Denmark, Germany, Greece, India, Italy, Spain, and Switzerland in the early 1990s. By 2007, at least 37 countries and 9 states/provinces had adopted such policies, more than half of which have been enacted since 2002 (Table 8, Table 9). Feed-in tariffs have clearly spurred innovation and increased interest and investment in many countries. These policies have had the largest effect on wind power, but have also influenced solar PV. Strong momentum for feed-in tariffs continues around the world as countries enact new feed-in policies or improve existing ones. For example, Portugal modified its feed-in tariff to account for technology differences, environmental impacts, and inflation. Austria amended its renewable electricity law to permit a new feed in tariff system. Spain modified feed-in tariff premiums (which are added to base power prices) to de-couple premiums from electricity prices and avoid windfall profits when electricity prices rose significantly. And Germany proposed modifications to its "EEG" ("Erneuerbare-Energien-Gesetz" – Renewable Energy Law) feed-in law. Elsewhere, Indonesia revised its feed-in tariff to cover plants up to 10 MW in size, from a previous limit of 1 MW. Thailand adopted a new feed-in policy for wind, solar, biomass, and micro-hydro. The Canadian province of Ontario enacted a feed-in tariff for a similar set of technologies. Nationally, Canada adopted a feed-in tariff premium which will provide CAD $0.01/kWh to renewables power projects constructed through 2011 and which is expected to cover an additional 4 GW of capacity. Many new feed-in tariffs directed specifically at solar PV appeared during 2006/2007.

Table 7: Renewable Energy Promotion Policies[1]

Country	Feed-in tariff	Renewable Portfolio Standard	Capital subsidies, grants, or rebates	Investment or other tax credits	Sales tax, energy tax, excise tax, or VAT reduction	Tradable renewable energy certificates	Energy production payments or tax credits	Net metering	Public investment, loans, or financing	Public competitive bidding
Developed and transition countries										
Australia		✓	✓			✓			✓	
Austria	✓		✓	✓		✓			✓	
Belgium		✓	✓		✓	✓		✓		
Canada	(*)	(*)	✓	✓	✓			(*)	✓	(*)
Croatia	✓			✓					✓	
Cyprus	✓		✓							
Czech Republic	✓		✓	✓	✓	✓		✓		
Denmark	✓				✓	✓		✓	✓	✓
Estonia	✓				✓					
Finland			✓		✓	✓	✓			
France	✓		✓	✓	✓	✓			✓	✓
Germany	✓		✓	✓	✓				✓	
Greece	✓		✓	✓						
Hungary	✓				✓	✓			✓	
Ireland	✓		✓	✓						✓
Italy	✓	✓	✓	✓		✓		✓		
Israel	✓									
Japan	(*)	✓	✓			✓		✓	✓	
Korea	✓		✓	✓	✓				✓	
Latvia	✓								✓	✓
Lithuania	✓		✓	✓					✓	
Luxembourg	✓		✓	✓						
Malta	✓				✓					
Netherlands	✓		✓	✓		✓	✓			
New Zealand			✓						✓	
Norway			✓	✓		✓				✓
Poland		✓	✓		✓				✓	✓
Portugal	✓		✓	✓	✓					
Romania					✓					
Russia			✓			✓				
Slovak Republic	✓			✓					✓	
Slovenia	✓								✓	
Spain	✓		✓	✓					✓	
Sweden		✓	✓	✓	✓	✓	✓			
Switzerland	✓		✓							
United Kingdom		✓	✓		✓	✓				
United States	(*)	(*)	✓	✓	(*)	(*)	✓	(*)	(*)	(*)

* Provincial/State policies but no national policies

Table 7: Renewable Energy Promotion Policies[1] *continued*

Country	Feed-in tariff	Renewable Portfolio Standard	Capital subsidies, grants, or rebates	Investment or other tax credits	Sales tax, energy tax, excise tax, or VAT reduction	Tradable renewable energy certificates	Energy production payments or tax credits	Net metering	Public investment, loans, or financing	Public competitive bidding
Developing countries										
Algeria	✓			✓	✓	✓				
Argentina	✓		✓	(*)	✓		✓			
Brazil	✓								✓	✓
Cambodia			✓							
Chile			✓							
China	✓		✓	✓	✓				✓	✓
Costa Rica	✓									
Ecuador	✓			✓						
Guatemala				✓	✓					
Honduras				✓	✓					
India	(*)	(*)	✓	✓	✓		✓		✓	✓
Indonesia	✓									
Mexico				✓				✓		
Morocco				✓						
Nicaragua	✓			✓	✓					
Panama							✓			
Philippines			✓	✓	✓				✓	
South Africa			✓							
Sri Lanka	✓									
Thailand	✓		✓						✓	✓
Tunisia			✓	✓						
Turkey	✓		✓							
Uganda	✓								✓	

* Provincial/State policies but no national policies

Table 8: Cumulative Number of Countries/States/Provinces Enacting Feed-in Policies[1]

Year	Cumulative Number	Countries/States/Provinces Added That Year
1978	1	United States
1990	2	Germany
1991	3	Switzerland
1992	4	Italy
1993	6	Denmark, India
1994	8	Spain, Greece
1997	9	Sri Lanka
1998	10	Sweden
1999	13	Portugal, Norway, Slovenia
2000	13	—
2001	15	France, Latvia
2002	21	Algeria, Austria, Brazil, Czech Republic, Indonesia, Lithuania
2003	28	Cyprus, Estonia, Hungary, South Korea, Slovak Republic, Maharashtra (India)
2004	34	Italy, Israel, Nicaragua, Prince Edward Island (Canada), Andhra Pradesh and Madhya Pradesh (India)
2005	41	Karnataka, Uttaranchal, and Uttar Pradesh (India); China; Turkey; Ecuador; Ireland
2006	44	Ontario (Canada), Argentina, Thailand
2007	46	South Australia (Australia), Croatia

Note: Cumulative number refers to number of jurisdictions that had enacted feed-in policies as of the given year. A few feed-in policies shown have been discontinued. Source: All available policy references, including the IEA on-line Global Renewable Energy Policies and Measures database and submissions from report contributors.

Table 9: Cumulative Number of Countries/States/Provinces Enacting RPS Policies[1]

Year	Cumulative Number	Countries/States/Provinces Added
1983	1	Iowa (USA)
1994	2	Minnesota (USA)
1996	3	Arizona (USA)
1997	6	Maine, Massachusetts, Nevada (USA)
1998	9	Connecticut, Pennsylvania, Wisconsin (USA)
1999	12	New Jersey, Texas (USA); Italy
2000	13	New Mexico (USA)
2001	15	Flanders (Belgium); Australia
2002	18	California (USA); Wallonia (Belgium); United Kingdom
2003	19	Japan; Sweden; Maharashtra (India)
2004	34	Colorado, Hawaii, Maryland, New York, Rhode Island (USA); Edward Island (Canada); Andhra Pradesh, Karnataka, Ma(
2005	38	District of Columbia, Delaware, Montana (USA); Gujarat (In

IV. Effect of Public Policy on Innovation

Johnstone, Hascic, and Popp obtained empirical results to indicate that public policy has had significant influence on the development of new technologies in the area of renewable energy[10]. Results suggest that policy instrument choice also matters. Their results indicate that (with respect to patent activity in renewable energy) the only significant policy instruments are taxes, obligations, and tradable renewable certificates (called TRCs or green-tags)[11].

V. Technology Policies For Long-Term Climate Targets

Policies that promote technical change in the context of climate change have been analyzed[12]. Recognizing four objectives: near-term abatement, institution building, technology development and the strengthening of proactive actors, the Kyoto protocol addresses all four. The protocol could most likely be met by an increased use of commercially available technologies, e.g. biomass, wind, solar thermal heating, increased energy efficiency, and natural gas instead of coal. However, beyond 2030 more advanced technologies such as solar PV will be needed for large-scale deployment to avoid serious climatic risks. Hence another strategy is needed. To complement the Kyoto Protocol and addresses long-term targets governments need to introduce incentives that support the process of bringing new technologies to the shelf. A portfolio of instruments, including RD&D (Research, Development, and Deployment) funding, support for industry network formation, niche market creation, and institutional adaptation needs to be implemented in a balanced way to foster new industries and set in motion a process of self-sustained growth, driven by dynamic learning and scale effects, where cost reductions generate market growth which, in turn, generate investments and learning that lead to further cost reductions. In addition, this process will increase the strength and numbers of proactive actors. The political power of the new industries will grow in comparison to the entrenched players and this will make it easier to adopt better climate targets in the future. The two-pronged strategy offers an evolutionary escape route from carbon lock-in.

Care should be taken not to make technology-fostering instruments too broad. Limiting the use of the flexible mechanisms in the Kyoto protocol is probably too blunt as a tool to bring more advanced technologies to the shelf. Similarly, there is a risk that green electricity certificates in the EU will only benefit technologies that are close to being competitive. If they are not complemented with more technology specific instruments, more advanced technologies will not be brought to the shelf.

It should be noted that the cost difference between the two types of instruments, economy wide price incentives and technology-fostering instruments is at least an order of magnitude. For instance, the annual cost of meeting the Kyoto protocol (with the United States onboard) is estimated at around hundreds of billions of dollars annually. This is 100 times the current annual public renewable energy RD&D expenditure in IEA (International Energy Agency) countries. Similarly, the cost of a program to buy down the cost of PV from $6/Wpeak to $1/Wpeak, which would make PV competitive with conventional electricity production, could peak at no more than $1/MWh of electricity in OECD countries (1% of the price of residential electricity).

VI. Energy Viability of Photovoltaic Systems and Policy Implications

Since PVSTs generate no emissions during operation, the environmental impacts are mainly associated with emissions generated during production of the technology (and the disposal of modules and cells at the end of their useful life). Many of these impacts are associated with the consumption of energy in the production process.

Alsema and Nieuwlaar have reviewed the energy viability of photovoltaic energy technology, that is, the question whether photovoltaic systems can generate sufficient energy *output* in comparison with the energy *input* required during production of the system components[13]. The energy viability was evaluated mainly in terms of the energy pay-back time (EPBT) (Fig. 11, Fig. 12). For a grid-connected PV system under a medium-high irradiation level of 1700 kWh/m^2 per year the EPBT was 2.5-3 years; for roof-top systems and almost 4 years for large, ground mounted systems. The share of module frames and module supports is quite significant but should decrease overtime. Also, over time the CO_2 emissions should decrease considerably (Fig. 13).

However, in some areas policy actions may be needed to realize this, for example by stimulating energy-efficient production processes for PV modules and by promoting system designs which require less materials and energy.

Fig. 11: Energy Payback Time (Year) of Present PV Systems[13]

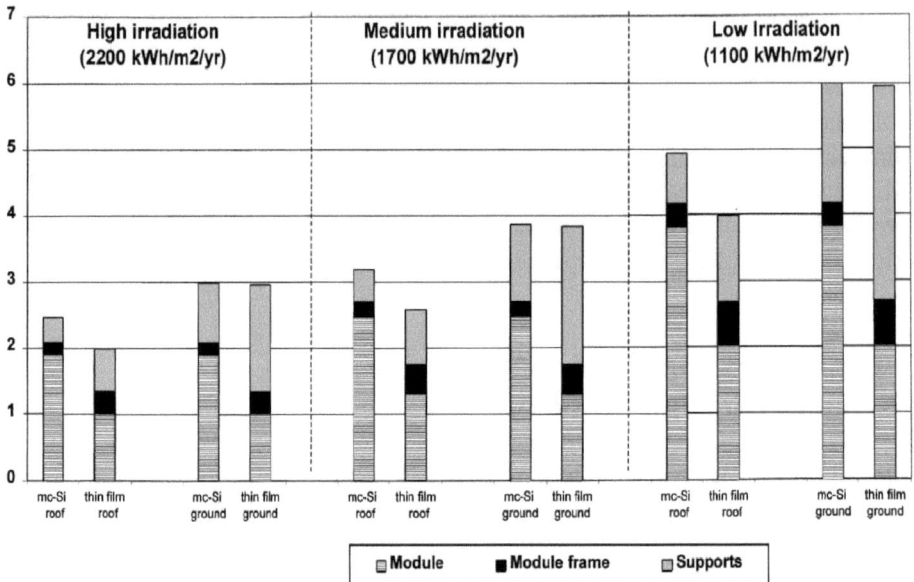

Fig. 12: Energy Payback Time (Year) of PV Systems[13]

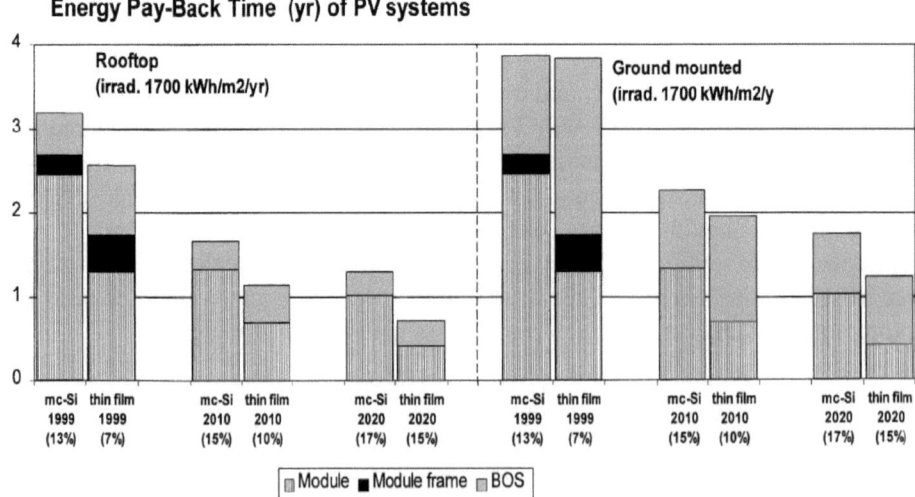

Fig. 13: CO_2 Emissions of Grid Supply Options (g/KWh [13]

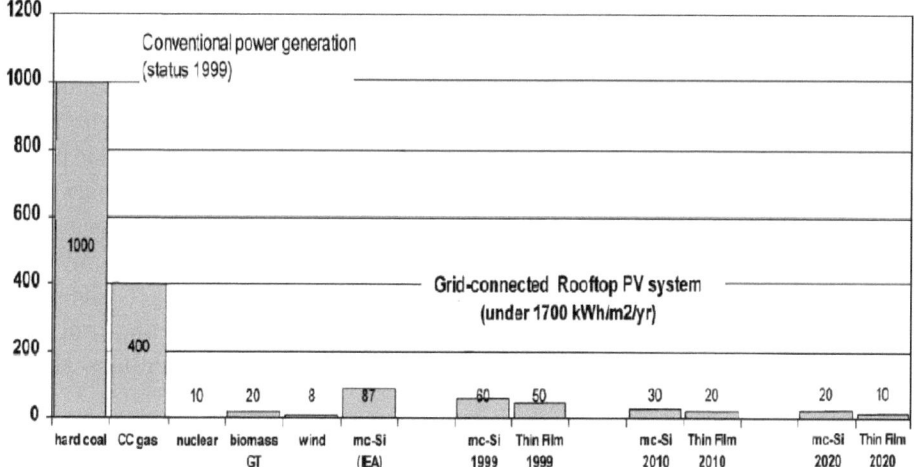

VII. Conclusion

There is a global trend to enact renewable energy policies and programs with a special interest in photovoltaic systems (PVSTs). PVST has universal appeal because of its technology roadmap, diverse deployment scenarios, and projected extremely low carbon footprint. The top five countries in solar PV deployment are: Germany, Spain, Japan, United States and South Korea. Government and private investments support renewables programs and the trend is growing. The policies act as catalysts for the industry and adopters for deployment of renewables. The investments act as a means to incubate and sustain the growth. The United States, Spain, China, and Germany are the investment leaders.

The renewable policies vary from country, but the top ones are:
- Feed-in tariffs
- Renewable portfolio standard (RPS)
- Capital subsidies or rebates
- Investments or other tax credits
- Sales tax, energy tax, excise tax, or VAT reduction
- Tradable renewable energy certificates (TRCs).

The policies are aligned to the national targets. By early-2009 renewable energy policy targets have been mandated by 73 countries. Typically, these targets are in the range of 5%-30% of the total power production. For example, the European

Union has established binding targets of 21% by 2010 and 34% by 2020. Even in countries such as the United States and Canada which do not have a national renewables policy, local state and provincial governments are taking charge and providing leadership in this area with an emphasis on solar PV power. PV is preferred because it can provide both wholesale and retail power (in commercial building and residential homes). Since local governments are more closely tied to PV deployment, usage, and environmental impact, their close engagement in policy is natural, productive, and can drive superior results. They also tend to enact more specific policies such as a PV policy versus a general renewable policy.

Although the focus is on near-term deployment, nations should also invest in PV technology innovation for 2030 and beyond. Performance targets should be included such as higher energy efficiency – by improving production processes and module design, energy viability, and environmental impact. Policies should be enacted to reflect these targets.

Technology transfer policies for PV can also expedite commercialization and deployment. These policies would include: ease of licensing new technologies, assistance with testing facilities, good system design that meets target users' needs, and easing licensing requirements for deployment. Currently, only nominal consideration has been given in this area.

Hence, the energy policy framework for PV should be expanded to include:

- Roadmap for PV technology innovation for 2030 and beyond.
- Technology transfer
- More engagement of local governments in technology innovation, transfer, and deployment.

In general, this framework can be extended to other renewables.

Concluding Remark: Historically, United States has opposed national renewables policies, but it appears that the new government administration of President Obama will reverse that direction.

VIII. References

[1] Renewable Energy Policy Network for the 21st Century (REN21) (2009). "Renewables Global Status Report 2009 Update".
[2] T. Daim (2009). "Accelerating the Adoption of Technologies for Meeting the Climate Change Challenges: The Case of Photovoltaic Solar Technology".
[3] Renewable Energy Policy Network for the 21st Century (REN21) (2007). "Renewables Global Status Report 2007".

[4] UNEP's Division of Technology, Industry and Economics (DTIE) under its Sustainable Energy Finance Initiative and New Energy Finance Limited (2008). "Global Trends in Sustainable Energy Investments".
[5] K. Cory, T. Couture and C. Kreycik (2009). "Feed-in Tariff Policy: Design, Implementation, and RPS Policy Interactions".
[6] Market capitalization statistics and analysis courtesy of New Energy Finance and Chris Greenwood.
[7] T. Bradford (2009). Prometheus Institute, e-mail to Janet Sawin, Worldwatch Institute.
[8] PV News, April 2009.
[9] J. Malaviya (2008). "On a Solar Mission: How India Is Becoming a Centre of PV Manufacturing".
[10] N. Johnstone, I. Hascic and D. Popp (2008). "Renewable Energy Policies and Technological Innovation: Evidence Based on Patent Counts".
[11] P. Bertoldi and T. Huld (2004). "Tradable certificates for renewable electricity and energy savings".
[12] B. Sanden and C. Azar. "Near-Term Technology Policies For Long-Term Climate Targets – Economy Wide Versus Technology Specific Approaches".
[13] E.A. Alsema and E. Nieuwlaar (2000). "Energy viability of photovoltaic systems".

Obstacles for Renewable Energy and Energy Efficiency in Chile
–
A Case Study from Hospitals

L. Muñoz del Campo[1]

Abstract

Over the last several years, Chile has been promoting the implementation of renewable energies and energy efficiency through a variety of initiatives, including a new electricity law. Despite the effort the country is making in promoting norms, regulations and incentives, the results of energy efficiency and renewable energy implementation are still well below forecasts and necessary to increase the strength of the Chilean energy supply.

However, all the country's efforts are geared towards one direction, notably to reduce the consumption of fossil fuels and increase renewable energy participation in the electricity matrix. Nevertheless, the results of implementing new ERNC generation still do not change the tendency or statistics of the country.

Through a practical approach to case studies from 2008 onward, I have realized that there were obstacles to implementing these types of programmes, which were generally a result of Chilean speech, practical decisions and a lack of a strategic vision in this matter.

I. Introduction

This document provides an example of the obstacles in implementing energy efficiency and renewable energies in Chile.

If we think about energy, we could point out that humankind has needed the energy associated with normal development since its existence – men need energy to live, perform activities, promote development, manage the environment and subsist. At first, this applied in general terms. Humankind has limited sources of energy available, including muscles and some animals. One person

---

1 munoz.lorena@gmail.com.

can produce 70 to 100 [W] for 6 to 8 hours or approx. 0.6 to 0.8 [kWh]. To do so, a person merely needs to eat about 3,000 to 3,500 [kcal], which will bring the efficiency of human energy production up to approx. 18 to 20%. An animal (i.e. a horse) can produce approx. 500 to 750 [W] (definition of HP) or 3 to 5 [kWh].

The invention of the vapour machine was the starting point for the industrial era at the end of the 19th century. During this period, human work was replaced by machine work and fossil fuel use, starting with wood and carbon[1]. The internal combustion engine was further developed throughout the 19th century. Initially, it was gas-operated, but its true development was achieved at the beginning of the 20th century with the improvement of Otto engine (gas-operated). The final impulse came with the advent of World War I and developments in aviation.

The use of electric power also began at the end of the 19th century. The creation of electric distribution networks quickly led to improvements in hydraulics, vapour turbines and industrial diesel engines. The industrial application of nuclear energy began in 1945.

Humankind has now increased its energy use. The beginning of the electrical era ushered in a new period during which humankind progressed from simple ways of life to modern times[2]. There is a slight unidirectional flow between employment and energy consumption and between energy consumption and GNP[3, 4].

The World Energy Council estimates that, over the next twenty years, world energy consumption will increase by approximately 50%, meaning that commercial energy will be able to be provided to 4000 million more users (2000 million which currently do not have it, plus another 2000 expected during this period)[7].

As we can see in Fig. 1[5], the OECD GNP increased at a rate higher than energy consumption from 1970 to 1990.

Therefore, it is clear that significant effort was made to create processes and energy-efficient systems during those years.[5]. As a matter of fact, we can point out renewable energy sources which are potentially capable of providing a significant portion of Europe's energy needs in the 21st century. The development of renewable energy is desirable for two main reasons. First, from the geopolitical standpoint, the future concentration of the world's oil resources in the Middle East includes the risks inherent in an increasing dependence on these resources. Secondly, there is widespread concern about the environmental effects of conventional energy consumption, in particular when it comes to global warming[6].

Fig. 1: Relation between Energy Consumption and GNP – OECD Countries[5]

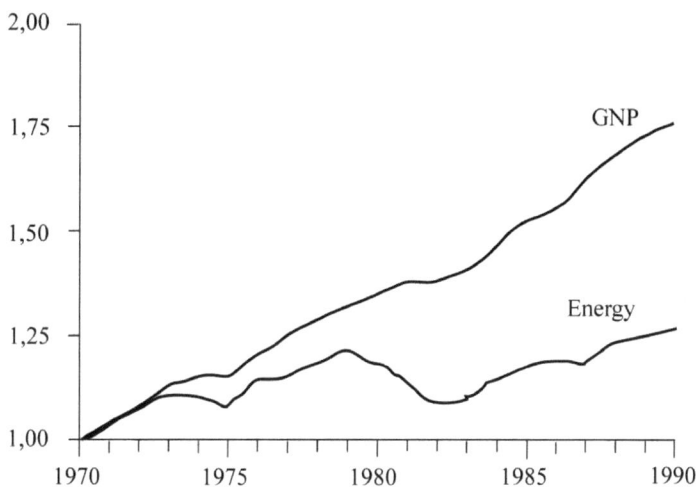

If we compare the evolution of GNP vs. energy consumption in OECD countries to the same factor in developing countries and Chile (Fig. 2 and 3[5]), a close correlation between GNP and energy consumption becomes clear.

Fig. 2: Relation between Energy Consumption and GNP – Developing Countries[5]

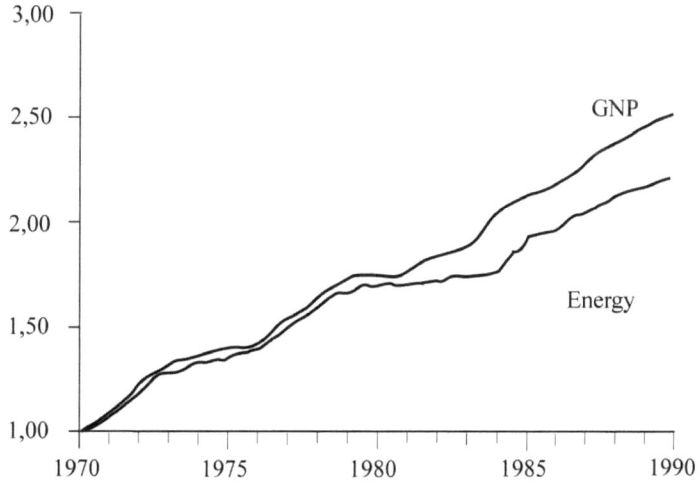

During this period, the GNP of the developing countries grew faster than the GNP in OECD countries (a factor 2 for developing countries versus 1.7 for

OECD countries). Nevertheless, the consumption of energy has increased at a rate faster than the GNP. This could be attributed to:

- Growth of infrastructure: almost all the developing countries should still carry out important efforts to improve their basic infrastructure and services.
- Basic processes: a significant portion of improving energy efficiency in OECD countries is a result of a strong "industrial conversion process".
- Incorporation of marginalized sectors: a great drama in developing countries is that large sectors of their population are out of the economic circuit. Incorporating these groups will require an increase in energy consumption. The issue of very marginal sectors of society must also be considered; there is a tendency is to initially associate energy consumption activities with poor efficiency.

Fig. 3: Relation between Energy Consumption and GNP – Chile[5]

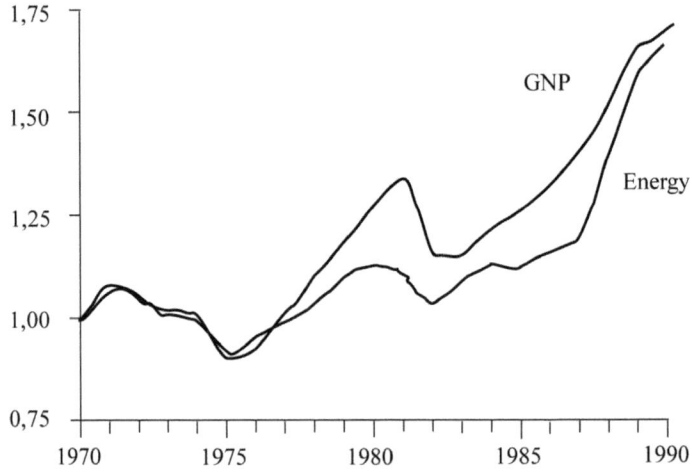

Figure 3[5] clearly shows the crisis of 1972-73 and the crisis of 1981 in Chile. It also highlights the fact that our GNP grew less from 1970 to 1990 than in developing countries.

If we move up to 2006, we can see that out of a total capacity of 12.326 MW in Chile, a mere 2.4% is provided by renewable energy. From 1982 to 2006, the total consumption of final energy increased more than three times (equal to an increase from 71,659 Tcalories to 240,579 Tcalories). An increase in demand by a factor of 3 is projected for the next 25 years. As shown in Fig. 4, in 2006, only 8% of total energy consumption was provided via hydraulic generation and 82% was provided by fossil fuels. Renewable energies do not appear to be at levels high enough to be included in the statistics[8].

Fig. 4: Energy flow for year 2006 in Chile (GWh)[8]

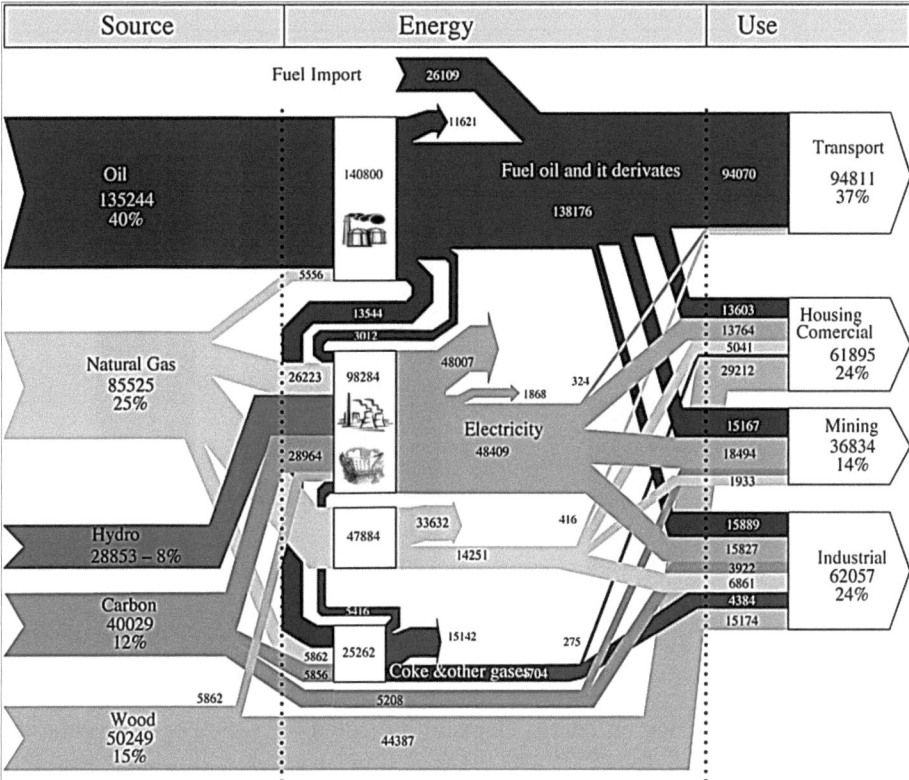

As in all Latin America countries, the decision-makers have been selecting fossil energy due to its apparent low cost[9]. As illustrated in Figure 5, this tendency in the electricity matrix implies an increase in the emission factor of the electricity grid in Chile.

Fig. 5: Emission of gases with global warming effect (tCO$_2$e/GWh) –Interconnected Electricity System – Chile [11]

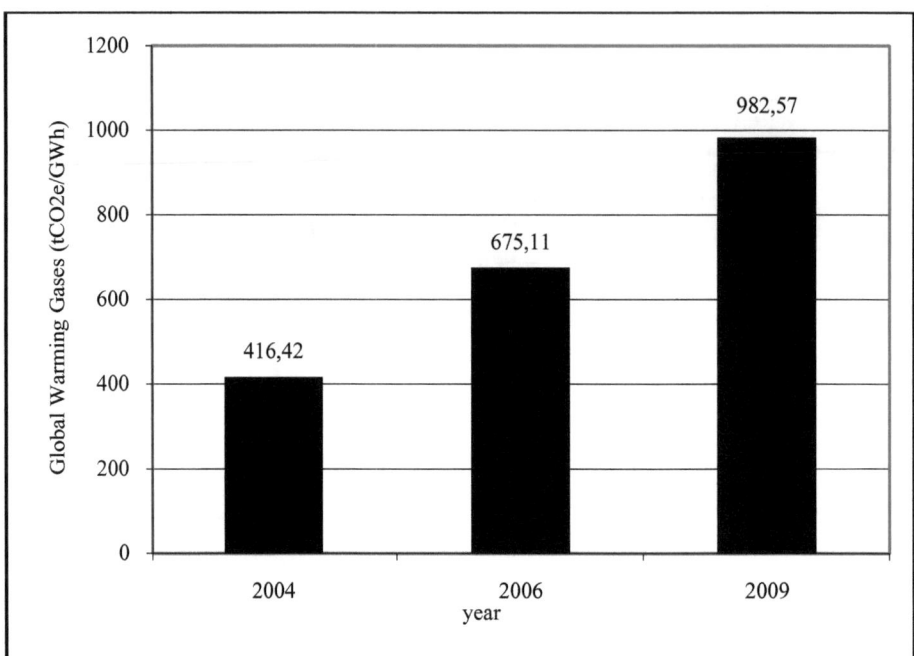

The significant increase in energy demand projected over the coming years, the importance of energy for GNP and the emission of gases with global warming effect has placed Chile in a good position to utilize energy efficiency measures and renewable energies to fulfil future demand.

Thus, the Chilean authorities executed the following measures to ensure security objectives and environmental sustainability in the energy sector:

- Promote the development of the ERNC via the support of private initiatives
- Perfect the legal framework to assure a non-discriminatory deal for the ERNC in the electric market
- Modify electric law to accelerate the development of the ERNC market in Chile.

Despite this fact, the numbers speak for themselves and the results are not as promising as they could be.

During last month of 2009 and the first quarter of 2010, I performed two different auditing procedures at hospitals with the sole objective of implementing energy efficiency and the renewable energy supply.

II. Methodology

The auditing procedure was established as described in ISO 50001, with each measure evaluated according to technical, economic and environmental perspectives.

The goals of the studies were as follows:

- Reduction in operating costs
- Increase in energy supply independence
- Simple and rapid implementation
- Sustainable development
- Reduced emission of gases with global warming effect
- Improve the institution's image.

Environmental issues were evaluated while considering the following points of view:

- Sustainability of the business
- Reduction of environmental liabilities
- Improving the possibility of emission of carbon credits to finance part of the energy efficiency programme.

The audited facilities were two hospitals in central Chile:

- Facility #1: hospital in the private sector
- Facility #2: hospital that belongs to a public institution
- Both hospitals have a similar period of operation (approx. 20 years)
- Both hospitals offer the same complexity of services.

For both facilities, we evaluated over two years of data in the following areas

- Energy consumption
- Use of industrial gases
- Direct energy generation
- Energy consumption for hot water production and vapour generation
- Both clients were provided with reports on the auditing procedure and energy efficiency management system.

Each report also included an economic evaluation and carbon strategy from project implementation to carbon credit emission.

III. Results

Both facilities were interested in energy efficiency implementation, initially over the mid-term, in order to implement renewable energies to supply energy to its facility. Both clients were keen to implement renewable energy; they were more than enthusiastic at the beginning of the auditing procedure.

One of the results of the auditing procedure was the need to implement two kinds of measures:

o Simple measures with high impact
o Measures related to concluding international treaties such as Montreal.

The aim of each set of measures is to reduce energy consumption, reduce cost and stabilize energy demand at a new level to enable the implementation of renewable energy. Technical, economic and environmental evaluations were conducted.

A. Facility #1 – Results

The followings projects were analyzed:

- 2010:
  o Chillier: shut down of two boilers – one natural gas boiler and one dual boiler (diesel – gas)
  o Replacement of lights.
- 2011:
  o R22 & R134A change to gases without ozone depletion effect and/or global warming effect.

The projects will have the following implications in terms of energy reduction:

o Natural gas: 452,000 m^3/year
o Diesel oil: 10,000 l/year
o Power supply: 80% reduction or 6.38 GWh.

Although energy efficiency will reduce running costs, it will also reduce global warming effect gases which could be translated into CERs (certified emissions reductions) shown in Table 1[12].

Table 1: Reduction in Global Warming effect gases[12]

Year	Reductions in global warming effect gases (tCO$_2$e/year)
2010	5,255
2011	32,064
2012	32,064
2013	32,064
2014	32,064
2015	32,064
2016	32,064
2017	32,064
2018	32,064
2019	32,064
Total	**293,830**

B. Facility #2

The projects analyzed were divided into the following categories:

- Power supply (Table 2)[13]
- Heat supply (Table 3)[13].

Table 2: Reduction in Power Consumption[13]

Measure proposed	% reduction	Amount reduced	Unit
Establish an organisation	1%	39,820	KWh/year
Programme for creating human resources	1%	39,820	KWh/year
Use of 2 x 36W, T5 fluorescent tubes with ballast (5000 app.)	12%	460,000	KWh/year
Technical management power factor	19%	768,000	KWh/year
Technical management of electrical charges	1%	39,820	KWh/year
	34%	1,347,460	KWh/year

Table 3: Reduction in Heat Supply[13]

Measure proposed	Reduction in natural gas consumption (m^3/year)
Change of windows – thermo panel	63,379
Install thermo panel door	99,627
Isolate areas	703,899
Isolate vapour installation	112,993
Eliminate R-12	234

Analyzed projects also indicate global warming effects gases:

- Power supply: 6309 tCO2e/year
- Heat supply: 15,650 tCO2e/year
- Total emission reduction per year 21,959 tCO2e/year
- During a ten-year project, emission reduction could be evaluated as 219,590 tCO2e.

As can be seen, both projects were comparable but the results were completely different in terms of implementation.

Facility #1 took the environmental issue to a strategic position by trying to increase the value creation through this variable.

Facility #2 decided not to move forward with this topic, even though it presents advantages to the baseline situation.

IV. Conclusion

Countries with emerging economies, based on the international commercialization of products, have two key points of concern:

- Economically secure and sufficient energy provision which demonstrates competence on global markets.
- To ensure access to global markets without restrictions, no taxes based on environmental or excessive obligation protectionism.

Chile is not the exception to these rules, but we can observe a dissociated point of view and dichotomy among decision- makers in Chile.

As pointed out by Mantilla[10], "Then, the focus is not the fulfilment, but the coherence (between targets and goals, between actions and programs, and the alignment according to). That's why the starting point at a managerial level is strategic planning and, as a consequence, how (long-term) strategies are lined up with (short-term) operations."

Based on the previous paragraph, it can be stated that it is the coherence process – including the financial/environmental variable as a fundamental component – which may explain, to a better extent, value generation within the scope of a knowledge economy.

On the one hand, we could see the lack of coherence as causing the organization make the more inconvenient decision related to energy matters, in addition to the vision they already have. On the other hand, coherence, good communication and a strong network are guiding the other facility into the implementation of an energy management system and a new way to evaluate projects.

This paper presents the lack of organizational skills as one of the obstacles in implementing energy efficiency and/or renewable energies. Another barrier appears to be a new connection between a relational approach to the environment and a new scope of analysis in both points of view – the associated financial risk and the relational approach – of an organization. The environmental status of the company itself does not indicate a reluctance to pursue this type of energy management system.

Nevertheless, if we include the organizational point of view, we can identify the probability of a company moving ahead with sustainable decision enhancements or reductions as a prediction of decision-making processes.

It could be interesting to evaluate this decision-making process, as related to organizational behaviour, while the auditing procedure is being performed in order to include it in developing a strategy to avoid obstacles to implementation.

V. References

[1] Brian Fagan (2008). El Gran Calentamiento. Ed. Gedisa S.A.
[2] Ramón Casilda Bejar (2008). "Energía y Desarrollo Económico en America Latina". Boletin de desarrollo económico 2750, pp. 32-44.
[3] S.H.Yu Eden and Been-Kwei Hwang (1984). "The relationship between energy and GNP: further results" Energy Economics, vol. 6, Issue 3, July 1984, pp. 186-190.
[4] E.S.H. Yu and J.Y. Choi (1985). "The Causal Relationship Between Energy and GNP: An International Comparison" Journal of Energy and Development, vol. 10, p. 2.
[5] Wolfgang Palz (1990). "Renewable Energy in Europe" International Journal of Sustainable Energy, 1478-646X, vol. 9, Issue 2, pp. 109-125.
[6] Emily A. Heaton, Stephen P. Long, Thomas B. Voigt, Michael B. Jones and John Clifton Brown, "Mitigation and Adaptation Strategies for Global Change", vol. 9, no. 4, pp. 433-451 (DOI: 10.1023/B:MITI.0000038848. 94134.be).
[7] Ramón Casilda Bejar (2002). "Energía y desarrollo económico en America Latina", Boletín Económico de ICE N°2750, 32 from 2 to 8 December 2002.
[8] Raúl O'Ryan (2008). PROGRAMA DE GESTIÓN Y ECONOMÍA AMBIENTAL (PROGEA) UNIVERSIDAD DE CHILE Departamento de Ingeniería Industrial Diseño de un Modelo de Proyección de Demanda Energética Global Nacional de Largo Plazo Informe Final Preparado para la Comisión Nacional de Energía – 30 June 2008.

[9] Jean Acquatella (2008). "Energia y cambio climático: oportunidades para una política energética integrada en América Latina y el Caribe". CEPAL & GTZ. Publicaciones de las Naciones Unidas, December 2008.

[10] Samuel Alberto Mantilla (1999). Capital Intelectual y Contabilidad del Conocimiento. SBN 10: 9586483665, ISBN 13/Cód Barra: 9789586483667, 2004, 3rd edition.

[11] Calculated using the "Tool to calculate the emission factor for an electricity system", Version 01.1, EB 35. This methodology determines (emission factor) in tons CO2 per GWh (tCO2/GWh) for Chile, based on the total number of facilities generating energy. cdm.unfccc.int/methodologies/.../tools/am-tool-07-v1.1.pdf.

[12] Lorena Muñoz del Campo and Alberto Learreta (2010). Final report "Estudio de Linea Base para la determinación de huella de carbono de HTS", February 2010.

[13] Lorena Muñoz del Campo and Alberto Learreta (2010). Final Report "Auditoría Energética a dos Reparticiones de la Armada de Chile ID Licitación: 3108-88-LE09 DIRECCIÓN DE PROGRAMAS, INVESTIGACIÓN Y DESARROLLO DE LA ARMADA DE CHILE", April 2010.

About the Authors

Ing. *Mariel Álvarez* is part of the supporting staff at the Regional Offices of the Energy Ministry for Atacama and Coquimbo, in Chile.

Facultad de Ciencias, Físicas y Matemáticas
Universidad de Chile
Av. Tupper 2007
Casilla 412-3
8370451 Santiago
Chile

Nelson Amaro is the Director of Guatemalan office of the CELA-Network for Climate Change Technology Transfer Centres in Europe and Latin America, a project financed by the Alfa III Program of the European Union. He is also an Adviser in International Cooperation for the Vicepresidency of Galileo University, in Guatemala.

Universidad de Galileo
Asesoría de Vicerrectoría y Cooperación Internacional
7a, Avenida, calle Dr. Eduardo Suger Cofiño, Zona 10,
Ciudad Guatemala
Guatemala CA
Contact e-mail: nelsonamaro@galileo.edu

Franziska Buch is a German economist, with a MSc in Latin American Studies, currently doing PhD studies at HAW Hamburg and London Metropolitan Business School. She works at the Bolivian Catholic University, as a manager of the projects JELARE and REGSA.

Instituto de Investigaciones Socio-Económicas (IISEC)
Universidad Católica Boliviana "San Pablo"
Av. 14 de Septiembre 4807 (Calle 2), Obrajes
La Paz
Bolivia
Contact e-mail: fbuch@ucb.edu.bo

Jacques Clerc
Facultad de Ciencias
Físicas y Matemáticas
Universidad de Chile

Av. Tupper 2007
Casilla 412-3
8370451 Santiago
Chile
Contact e-mail: jaclerc@dii.uchile.cl

Dr. *Tugrul Daim* is an associate professor and the director of the PhD Program in Technology Management at Portland State University, Portland Oregon. He has over 20 years of experience in industry, academia and the government sectors.

Portland State University
Department of Engineering and Technology Management
Portland OR 97201
USA
(503) 725 4582
Contact e-mail: tugrul@etm.pdx.edu

Guilherme Dantas is a Ph.D. candidate at the Federal University of Rio de Janeiro (UFRJ), Brazil. He is also Senior Researcher at the Group of Studies of Eletric Energy (GESEL).

Federal University of Rio de Janeiro – UFRJ
Group of Studies of Electric Energy (GESEL)
Instituto de Economia / UFRJ
Av. Pasteur, 250 / 2°andar / 226
Urca - Rio de Janeiro/RJ
CEP: 22290-240
Contact e-mail: guilhermecrvg@yahoo.com.br

Professor *Nivalde de Castro* is the Head of the Group of Studies of Electric Energy (GESEL) at the Federal University of Rio de Janeiro (UFRJ), Brazil.

Federal University of Rio de Janeiro – UFRJ
Group of Studies of Electric Energy (GESEL)
Instituto de Economia / UFRJ
Av. Pasteur, 250 / 2°andar / 226
Urca - Rio de Janeiro/RJ
CEP: 22290-240
Contact e-mail: nivalde@yahoo.com

Mr. *Manuel Díaz* is Director of the Programme for Environmental Economics and Management (PROGEA) and researcher at the Industrial Engineering Department, Universidad de Chile. he is a project consultant and researcher in Environmental Management and Energy, Mining, Climate Change, Risk Analysis, Environmental Economics and Information Systems.

Director
Program for Environmental Economics and Management (PROGEA)
Fundación para la Transferencia Tecnológica
Beauchef 993
Santiago
Chile
Contact e-mail: mdiazr@dii.uchile.cl

Facultad de Ciencias, Físicas y Matemáticas
Universidad de Chile
Av. Tupper 2007
Casilla 412-3
8370451 Santiago
Chile
Contact e-mail: mdiazr@dii.uchile.cl

Francisco Espín Sánchez received his Bachelor of Electrical Engineering at the Universidad Politécnica de Cartagena (Spain). Hi is currently with CEO and CTO, Gehrlicher Solar Spain.

Francisco Espín Sánchez
Gehrlicher Solar Spain
c/ Valle Guadalentin, 2, 2°
30180 Bullas (Murcia)
Spain
Contact e-mail: francisco.espin@gehrlicher.com

Rodrigo García Palma is a civil engineer, with a Diploma on Environmental Engineering of the Pontificia Universidad Católica de Chile. He has 9 years of experience in public institutions and international private companies, in the field of sustainable development. He currently serves as Technical Manager of Renewable Energy Center – CORFO. In this role he has been able to design new incentives for the promotion of renewable energy in Chile. He has led a team that delivers inputs for policy makers and private entrepreneurs. At the same time, he has coordinated various consultancies in the areas of renewable energy, carbon markets and technology transfer for climate change.

Technical Manager
Centro de Energias Renovables
Santiago de Chile, Chile
Contact e-mail: rsgarciapalma@gmail.com

Natalia Garrido is a geographer and is involved with the Progea at University of Chile.

Facultad de Ciencias, Físicas y Matemáticas
Universidad de Chile
Av. Tupper 2007

Casilla 412-3
8370451 Santiago, Chile
Contact e-mail: nati.garrido@gmail.com

Dr. *Emilio Gómez Lázaro* was born in 1969. He received the Electrical Engineering and Ph.D degrees in the Universidad Politécnica de Valencia, Spain, in 1995 and 2000, respectively. Currently, he is an Associate Professor at the University of Castilla-La Mancha, being the head of the Renewable Energy Research Institute. His main interest are design and modelling renewable based power plants, mainly wind farms and wind turbines, and power system operations with large amounts of renewable based power. He is a senior member of the IEEE.

Renewable Energy Research Institute
DIEEAC, EDII-AB
Universidad de Castilla-La Mancha
Campus Universitario. Avda. de España, s/n
02071 Albacete
Spain
Contact e-mail: emilio.gomez@uclm.es

Julia Gottwald works at the Hamburg University of Applied Science in Germany in the Research and Transfer Centre "Application of Life Sciences". She is coordinating several EU funded projects on technology transfer and development cooperation in the renewable energy sector in Latin America as well as African, Pacific and Caribbean small island development states.

Research and Transfer Centre "Applications of Life Sciences"
Faculty of Life Sciences
Hamburg University of Applied Sciences
Lohbruegger Kirchstraße 65
21033 Hamburg
Germany
Contact e-mail: Julia.gottwald@ls.haw-hamburg.de

Javier Guerrero Pérez received his Bachelor of Electrical Engineering in 2004, and his Master Degree in 2007 at the Universidad Politécnica de Cartagena (Spain). He is currently working with Gehrlicher Solar Spain at the Ground Mounted System Department.

c/ Valle Guadalentin, 2, 2°
30180 Bullas (Murcia)
Spain
Contact e-mail: Javier.Guerrero@gehrlicher.com

About the Authors

Dr. *Guillermo Jiménez* is part of the Research and Development team of the FCFM – Energy Center at Universidad de Chile.

Facultad de Ciencias, Físicas y Matemáticas
Universidad de Chile
Av. Tupper 2007
Casilla 412-3
8370451 Santiago
Chile
Contact e-mail: gjimenez@ing.uchile.cl

Pratima Khadoo-Jeetah is an academic staff in the department of Chemical and Environmental Engineering. She is also doing her PhD on bioethanol.

Lecturer
Chemical and Environmental Engineering Department
University of Mauritius
Reduit
Mauritius
Contact e-mail: p.jeetah@uom.ac.mu

I. Kliopova is Associate professor at Kaunas University of Technology (KTU), Lithuania.

Institute of Environmental Engineering (APINI)
Kaunas University of Technology
K. Donelaičio str. 20
LT-44239 Kaunas
Lithuania
Contact e-mail: irina.kliopova@ktu.lt

Professor *Walter Leal Filho* is the Head of the Research and Transfer Centre "Applications of Life Sciences" at the Hamburg University of Applied Sciences, in Hamburg, Germany

Walter Leal Filho
Research and Transfer Centre "Applications of Life Sciences"
Faculty of Life Sciences
Hamburg University of Applied Sciences
Lohbruegger Kirchstraße 65
21033 Hamburg
Germany
Contact e-mail: Walter.Leal@haw-hamburg.de

André Leite is Professor at the Federal University of Fronteira Sul (UFSS) in Brazil. He is also a Senior Researcher at the Group of Studies of Eletric Energy (GESEL).

Federal University of Fronteira Sul – UFFS
Av. Getúlio Vargas, 609N, 2o andar,
Chapecó /SC
89812-000
Contact e-mail: andre_leite@hotmail.com

Joaquin Martínez Jiménez received his Bachelor of Electrical Engineering Degree at the Universidad Politécnica de Cartagena (Spain). Hi is currently Technical Director of Gehrlicher Solar Spain.

Gehrlicher Solar Spain
c/ Valle Guadalentin, 2, 2º
30180 Bullas (Murcia)
Spain
Contact e-mail: joaquin.martinez@gehrlicher.com

María Mendiluce, economist with a PhD degree in engineering by the Spanish Universidad de Comillas (Madrid), works in the World Business for Council Sustainable Development in Geneva, where she is leading several projects in the area of energy and climate change. In addition she is a researcher and has written several publications around the area of technology transfer, carbon pricing and Spanish energy demand and transportation.

Manager Energy & Climate in the World Business Council for Sustainable Development
Eco Patent Commons and Energy & Climate in the World Business Council for Sustainable Development
Geneva
Switzerland
Contact e-mail: Mendiluce@wbcsd.org

Professor *Romeela Mohee* is presently the Dean of the Faculty of Engineering. She is basically an Energy Engineer and she has 20 years of experience in waste management. She has one of the biggest research in Environmental Engineering in Mauritius.

Dean of Faculty of Engineering
University of Mauritius
Reduit
Mauritius
Contact e-mail: rmh@uom.ac.mu

Dr. *Angel Molina-Garcia* received his Electrical Engineering Degree at the Universidad Politécnica de Valencia (Spain) in 1998, and his PhD at the Universidad Politécnica de Cartagena (Spain) in 2003. He is currently Associate Professor, Dept. of Electrical Eng, Universidad Politécnica de

Cartagena (Spain). His research interests include renewable energy resources, distributed generation and energy efficiency.

Dept. of Electrical Engineering Degree
Universidad Politecnica de Cartagena
30202 Cartagena
Spain
Contact e-mail: angel.molina@upct.es

Dario Morales received the Electrical Engineer degree from the University of Concepción, Chile and the Ph.D. degree from the University of Paris XI and l'École Supérieure d'Électricité-Supélec, Paris, France. Dr. Morales is currently the Energy Coordinator of InnovaChile which is a governmental agency dedicated to support to Chilean firms to improve their competitiveness in national and international markets by promoting the development of innovative process. Its scope of action ranges from individual companies and networked firms to full production chains, including clusters or geographic groups of companies working in a particular industry.

Energy Coordinator of InnovaChile
CORFO
Santiago de Chile
Chile
Contact e-mail: dmorales@corfo.cl

Lorena Muñoz del Campo is a Chilean Biochemist, Business Diploma and Master in Finance. She has almost 20 years experience in Environmental Management. During the last years, she has undertaken CDM projects from the conception, implementation, until the emission of carbon credit. She has also contributed to the development of a new methodology towars the implementation of new emissions reductions and finance structures for these kind of projects. At present, she advises organisations on strategic issues, specifically searching, creating and establishing strategies and plans that allow them to create economic value from environmental resources.

Universidad de Viña del Mar
Escuela de Ingenieria Ambiental
Área Química Ambiental / Bonos de Carbono
Avenida Santa Luisa – 395
2541206 Vina Del Mar
Chile
Contact e-mail: munoz.lorena@gmail.com
www.lorenamunoz.cl

Raúl O'Ryan

> Facultad de Ciencias, Físicas y Matemáticas
> Universidad de Chile
> Av. Tupper 2007
> Casilla 412-3
> 8370451 Santiago
> Chile
> Contact e-mail: roryan@dii.uchile.cl

V. Petraškienė is PhD student at the Institute of Environmental Engineering, KTU, Lithuania.

> Institute of Environmental Engineering (APINI)
> Kaunas University of Technology
> K. Donelaičio str. 20
> LT-44239 Kaunas
> Lithuania
> Contact e-mail: vpetraskiene@gmail.com

Ana Pueyo is a PhD candidate in Industrial Engineering by the Technical University of Madrid and is trained in management and economics by the London School of Economics and the Universidad Autonoma de Madrid. She is a Research Fellow at the Climate Change team of the Institute of Development Studies of the University of Sussex, UK. She has done extensive research about clean energy technology transfer to developing countries and has managed several projects related to energy and climate change policy for the European Commission, the UK Government, the Spanish Government and several NGOs and private companies.

> Institute of Development Studies – University of Sussex
> Library Rd
> Falmer, Brighton, East Sussex BN1 9RE
> United Kingdom
> Contact e-mail: anapueyo@hotmail.com

José Baltazar Salguirinho Osório de Andrade Guarre has a degree in Economics, a MSc in Social and Economic Development and a Doctorate in Political Sciences/International Relations. He is presently the Dean of the Business School at the UNISIL, the University of the South if Santa Catarina, in Brazil.

> Universidad do Sul de Santa Catarina
> RuaTrajano 219, Centro
> 88010-010 Florianóplis – SC
> Brazil
> Contact e-mail: baltazar.guerra@unisul.br

About the Authors

Veronika Schulte is biologist and works at the Hamburg University of Applied Science in Germany in the Research and Transfer Centre "Applications of Life Sciences". She is coordinating several EU funded projects on technology transfer and development cooperation in the renewable energy sector in Latin America as well as African, Pacific and Caribbean small island development states.

Research and Transfer Centre "Applications of Life Sciences"
Faculty of Life Sciences
Hamburg University of Applied Sciences
Lohbruegger Kirchstraße 65
21033 Hamburg
Germany
Contact e-mail: veronika.schulte@haw-hamburg.de

Nasir J. Sheikh is a doctoral student at the PhD Program in Technology Management at Portland State University, Portland Oregon. He has over 30 years of experience in industry.

Portland State University
Department of Engineering and Technology Management
Portland OR 97201
USA
(503) 725 4660
Contact e-mail: nasirsheikh5@gmail.com

Professor *J.K. Staniškis* is Director of the Institute of Environmental Engineering, at Kaunas University of Technology (KTU), Lithuania.

Institute of Environmental Engineering (APINI)
Kaunas University of Technology
K. Donelaičio str. 20
LT-44239 Kaunas
Lithuania
Contact e-mail: jurgis.staniskis@ktu.lt

Dinesh Surroop is currently working as Lecturer at the Department of Chemical and Environmental Engineering, Faculty of Engineering, University of Mauritius, Mauritius. His fields of specialization are waste management, energy from waste and energy management.

Department of Chemical and Environmental Engineering
Faculty of Engineering
University of Mauritius
Reduit
Mauritius
Contact e-mail: d.surroop@uom.ac.mu

Thematic Index

abatement costs 105–26

biodegradable waste 145–58

Chile 69–87
clean energy 11–39
cleaner production 145–58
climate change 171–81, 183–211
Club of Rome 11–39
competencies 41–67
competitiveness 127–42
costs 127–42
curriculum modernization 41–67

dependence theories 11–39
development paradigms 11–39
disturbances 159–69

economic development 183–211
electricity sector 127–42
energy 183–211
energy and development 183–211
energy consumption 105–26
energy efficiency 183–211
energy policy 127–42, 171–81
energy recovery 145–58
environment 11–39, 41–67

grain processing 145–58
Greenhouse gases 105–26
grid codes 159–69

JELARE 41–67

knowledge transfer 89–104

labour market survey 89–104

mass and energy balance 145–58
mitigation options 105–26
modernisation 11–39
multidisciplinary 41–67

natural resources 183–211
non-BRIC countries 69–87

Organisation of Petroleum
   Exporting Countries, (OPEC)
   11–39

photovoltaic power plants 159–69
photovoltaics 171–81
Pilot module 41–67

real measurements 159–69
renewable energy 11–39, 41–67,
   69–87, 89–104, 171–81, 183–
   211

small developing island states 89–104
solar energy 171–81
solar technology 171–81
solid recovered fuel 145–58

student profile 41–67
sustainable development 11–39, 183–211

technology transfer 89–104
technology transfers 69–87

UNFCCC 69–87

voltage dips 159–69

wind industry 69–87
wind power 127–42
world systems theories 11–39